It's Been
Four Billion Years

The story of
Life on Earth
a million years
at a time

by jwcarvin

Printed in the United States of America

ISBN 978-0-9768183-9-7

The timeline
used in this book
adheres to a consistent
scale in an effort to convey
time realistically and proportionately
during the history of life on Earth,
from four billion years ago
to the present.

1. ORIGINS

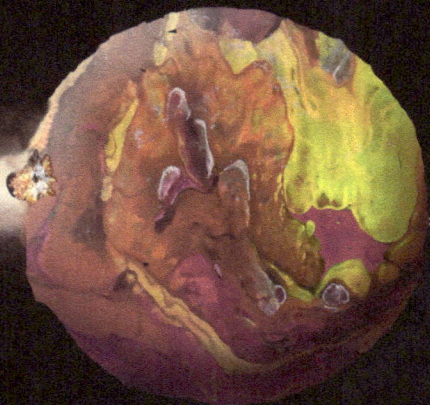

The eon named for Hell, when asteroids bombarded infant Earth, volcanoes spewed out noxious chemicals, and molten rock flowed everywhere, is finally at an end.

The Earth has cooled at last, and elements —
those building blocks created by the stars —
begin to form into the things we are.

With H_2O no longer boiled away,
the surface of the Earth's a vast expanse
of water, full of CO_2, with no
free oxygen to breathe, not even in
the atmosphere above. But heat inside
Earth's core erupts through hydrothermal vents
and forces molecules through porous rock,
which breaks up bonds and makes electrons jump
from molecule to molecule, and makes
the chemicals from which our lives are made.

H_2S

CO_2

H_2O

CO_2

NH_3

H_2O

H_2O

CO_2

CH_4

3,980,000,000
3,979,000,000
3,978,000,000
3,977,000,000
3,976,000,000
3,975,000,000
3,974,000,000
3,973,000,000
3,972,000,000
3,971,000,000
3,970,000,000
3,969,000,000
3,968,000,000
3,967,000,000
3,966,000,000
3,965,000,000
3,964,000,000
3,963,000,000
3,962,000,000
3,961,000,000

H_2

H_2

H_2S

CO_2

CH_4

H_2S

H_2O

PO_4

NH_3

N_2O

H_2O

N_2

CO_2

H_2

H_2O

H_2S

H_2S

H_2

H_2

H_2S

H_2

CO_2

NH_3

CO_2

H_2S

H_2

H_2

CO_2

H_2O

H_2O

CO_2

3,960,000,000
3,959,000,000
3,958,000,000
3,957,000,000
3,956,000,000
3,955,000,000
3,954,000,000
3,953,000,000
3,952,000,000
3,951,000,000
3,950,000,000
3,949,000,000
3,948,000,000
3,947,000,000
3,946,000,000
3,945,000,000
3,944,000,000
3,943,000,000
3,942,000,000
3,941,000,000

H₂

H₂S

H₂

CO₂

H₂O

H₂S

No fire below nor ultraviolet light
that inundates the Earth with deadly rays
can kill us off, as long as we remain
no more complex than simple chemicals.

CO2

H₂

H₂S

H₂

H₂O

H₂

H₂S

H₂O

CO₂

H₂

H₂

H₂O

3,940,000,000
3,939,000,000
3,938,000,000
3,937,000,000
3,936,000,000
3,935,000,000
3,934,000,000
3,933,000,000
3,932,000,000
3,931,000,000
3,930,000,000
3,929,000,000
3,928,000,000
3,927,000,000
3,926,000,000
3,925,000,000
3,924,000,000
3,923,000,000
3,922,000,000
3,921,000,000

H₂

H₂   H₂S

CO₂   H₂O

H₂S

CO₂

H₂O

H₂   PO₄   H₂S

CO2   H₂

H₂S   H₂   C₅H₄N₄

H₂S

H₂S   H₂

H₂   CO₂   H₂O

H₂S   H₂

H₂   CO₂   H₂O

3,920,000,000
3,919,000,000
3,918,000,000
3,917,000,000
3,916,000,000
3,915,000,000
3,914,000,000
3,913,000,000
3,912,000,000
3,911,000,000
3,910,000,000
3,909,000,000
3,908,000,000
3,907,000,000
3,906,000,000
3,905,000,000
3,904,000,000
3,903,000,000
3,902,000,000
3,901,000,000

As time goes on, among the molecules
that form from hydrogen and nitrogen,
from oxygen and carbon, are a few
we call the purines and pyrimidines,
which bond with sugars and with phosphate groups
to make nucleotides, which form long chains
that (someday) we will give the fancy name
"ribonucleic acid" (RNA).

$C_4H_4N_2$

$C_5H_4N_4$

$C_5H_5N_5$

$C_5H_5N_5O$

$C_5H_5N_5O$

$C_5H_5N_2O_2$

$C_4H_4N_2O_2$

PO_4

$C_5H_{10}O_5$

3,900,000,000
3,899,000,000
3,898,000,000
3,897,000,000
3,896,000,000
3,895,000,000
3,894,000,000
3,893,000,000
3,892,000,000
3,891,000,000
3,890,000,000
3,889,000,000
3,888,000,000
3,887,000,000
3,886,000,000
3,885,000,000
3,884,000,000
3,883,000,000
3,882,000,000
3,881,000,000

These purines and pyrimidines act like some sort of secret code, determining what happens to the chemicals nearby. They tell them how to rearrange themselves, assembling new proteins from smaller things, at times creating copies of themselves. They "replicate," and so it's said they live!

1,000,000 X Mag.

PO_4 — $C_5H_5N_5$
$C_5H_{10}O_5$
PO_4 — $C_4H_4N_2O_2$
$C_5H_{10}O_5$
PO_4 — $C_5H_5N_5O$
$C_5H_{10}O_5$
PO_4 — $C_4H_5N_3O$
$C_5H_{10}O_5$
PO_4 — $C_4H_4N_2O_2$
$C_5H_{10}O_5$ — $C_4H_5N_3O$
PO_4
$C_5H_{10}O_5$ — $C_5H_5N_5O$
PO_4
$C_5H_{10}O_5$ — $C_5H_5N_5$
PO_4
$C_5H_{10}O_5$ — C_5H_5N
PO_4 — $C_5H_5N_5$
$C_5H_{10}O_5$

PO_4 — $C_5H_5N_5$
$C_5H_{10}O_5$
PO_4 — $C_4H_4N_2O_2$
$C_5H_{10}O_5$
PO_4 — $C_5H_5N_5O$
$C_5H_{10}O_5$
PO_4 — $C_4H_5N_3O$
$C_5H_{10}O_5$
PO_4 — $C_4H_4N_2O_2$
$C_5H_{10}O_5$ — $C_4H_5N_3O$
PO_4
$C_5H_{10}O_5$ — $C_4H_5N_5O$
PO_4
$C_5H_{10}O_5$ — $C_5H_5N_5$
PO_4
$C_5H_{10}O_5$

3,860,000,000
3,859,000,000
3,858,000,000
3,857,000,000
3,856,000,000
3,855,000,000
3,854,000,000
3,853,000,000
3,852,000,000
3,851,000,000
3,850,000,000
3,849,000,000
3,848,000,000
3,847,000,000
3,846,000,000
3,845,000,000
3,844,000,000
3,843,000,000
3,842,000,000
3,841,000,000

Now, sometimes lengthy chains can break apart.
(The lengthier they get, the more they tend
to do just that.) But when they do, they're apt
to bond again, to rearrange themselves.
By breaking and rejoining, they evolve.

3,840,000,000
3,839,000,000
3,838,000,000
3,837,000,000
3,836,000,000
3,835,000,000
3,834,000,000
3,833,000,000
3,832,000,000
3,831,000,000
3,830,000,000
3,829,000,000
3,828,000,000
3,827,000,000
3,826,000,000
3,825,000,000
3,824,000,000
3,823,000,000
3,822,000,000
3,821,000,000

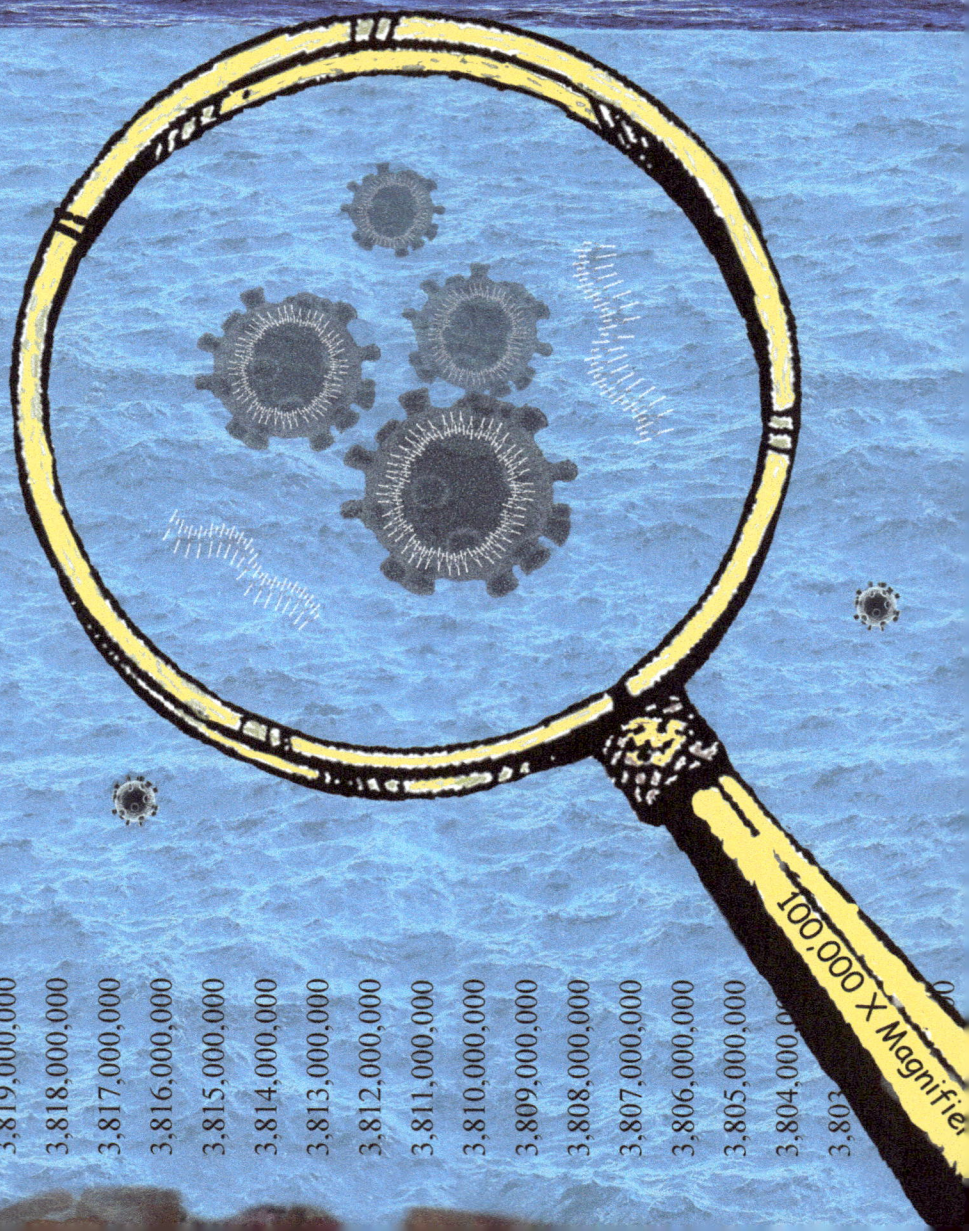

3,820,000,000 3,819,000,000 3,818,000,000 3,817,000,000 3,816,000,000 3,815,000,000 3,814,000,000 3,813,000,000 3,812,000,000 3,811,000,000 3,810,000,000 3,809,000,000 3,808,000,000 3,807,000,000 3,806,000,000 3,805,000,000 3,804,000,000 3,803

100,000 X Magnifier

Among these chains of RNA are some
called viruses that make hard protein shells
to serve as housings for themselves, then do
no more; they can't metabolize, and so
can't replicate, unless they break into
some other cells to steal their energy.

3,800,000,000
3,799,000,000
3,798,000,000
3,797,000,000
3,796,000,000
3,795,000,000
3,794,000,000
3,793,000,000
3,792,000,000
3,791,000,000
3,790,000,000
3,789,000,000
3,788,000,000
3,787,000,000
3,786,000,000
3,785,000,000
3,784,000,000
3,783,000,000
3,782,000,000
3,781,000,000

Now meanwhile, fatty acids, glycerols
and sterols form around some RNA,
becoming natural lipid barriers
that separate what's inside from the seas
of foreign things that circulate outside.

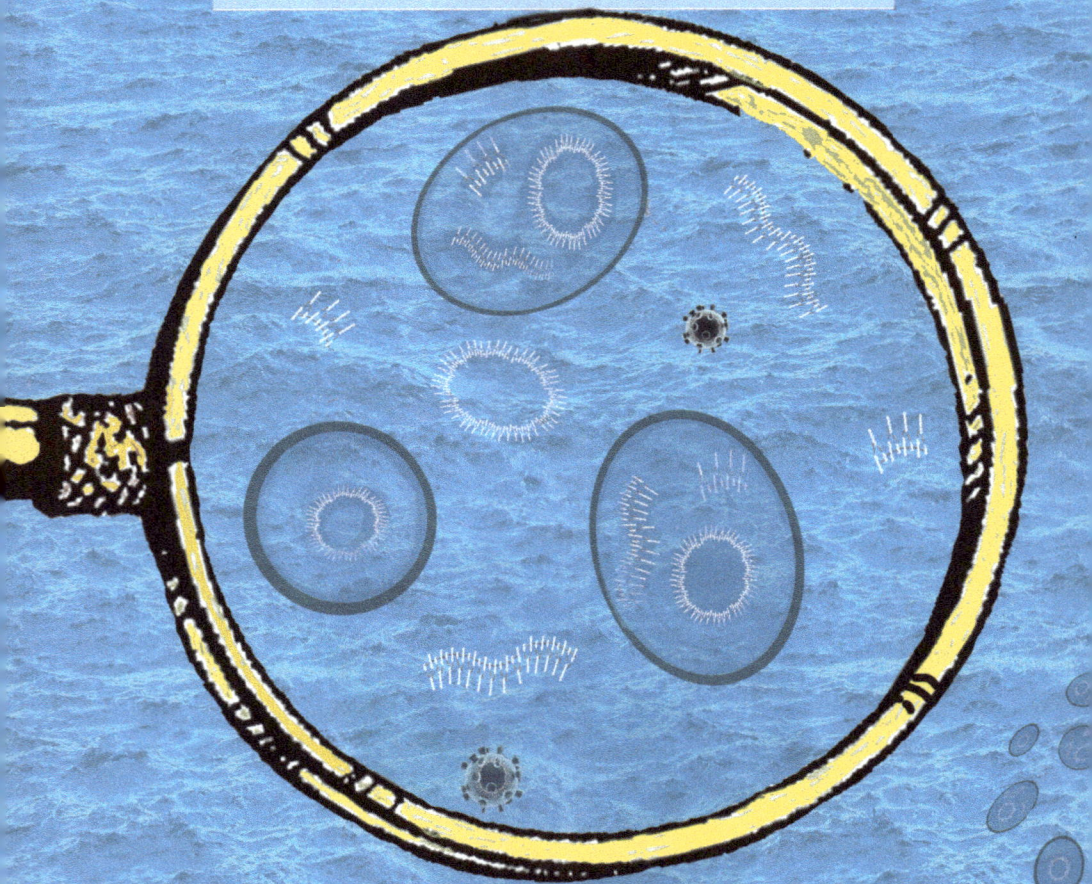

3,780,000,000
3,779,000,000
3,778,000,000
3,777,000,000
3,776,000,000
3,775,000,000
3,774,000,000
3,773,000,000
3,772,000,000
3,771,000,000
3,770,000,000
3,769,000,000
3,768,000,000
3,767,000,000
3,766,000,000
3,765,000,000
3,764,000,000
3,763,000,000
3,762,000,000
3,761,000,000

We call the cells they form prokaryotes.
They're tiny things, about a micron long.
It takes some hundred million years or more,
but once they learn to live inside those walls,
they're free to leave their thermal vents and go
explore the farthest reaches of the Earth.

3,760,000,000
3,759,000,000
3,758,000,000
3,757,000,000
3,756,000,000
3,755,000,000
3,754,000,000
3,753,000,000
3,752,000,000
3,751,000,000
3,750,000,000
3,749,000,000
3,748,000,000
3,747,000,000
3,746,000,000
3,745,000,000
3,744,000,000
3,743,000,000
3,742,000,000
3,741,000,000

2. MUTATION

Now, these long chains by which we replicate
contain the formulas for who we are
and how we work. They can't pass through our walls,
but smaller pieces do so easily,
and so we share short lengths of RNA,
incorporating them in our own chains.
It's evolution that protects us all.

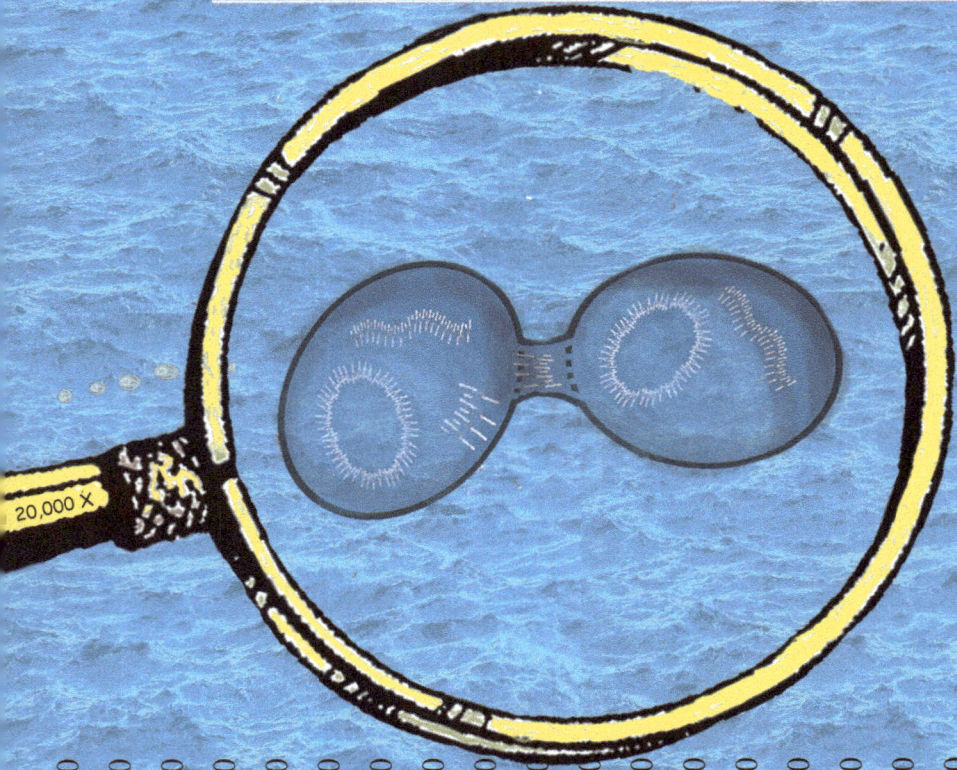

20,000 X

3,740,000,000
3,739,000,000
3,738,000,000
3,737,000,000
3,736,000,000
3,735,000,000
3,734,000,000
3,733,000,000
3,732,000,000
3,731,000,000
3,730,000,000
3,729,000,000
3,728,000,000
3,727,000,000
3,726,000,000
3,725,000,000
3,724,000,000
3,723,000,000
3,722,000,000
3,721,000,000

This all takes time, of course: a million years...
ten million more...   A book of life can be
a long, long read. How long before you'll want
to turn this page? A minute? Humor us:
wait forty million years before you do,
and feel the pace at which our lives evolve.

3,720,000,000
3,719,000,000
3,718,000,000
3,717,000,000
3,716,000,000
3,715,000,000
3,714,000,000
3,713,000,000
3,712,000,000
3,711,000,000
3,710,000,000
3,709,000,000
3,708,000,000
3,707,000,000
3,706,000,000
3,705,000,000
3,704,000,000
3,703,000,000
3,702,000,000
3,701,000,000

3,700,000,000
3,699,000,000
3,698,000,000
3,697,000,000
3,696,000,000
3,695,000,000
3,694,000,000
3,693,000,000
3,692,000,000
3,691,000,000
3,690,000,000
3,689,000,000
3,688,000,000
3,687,000,000
3,686,000,000
3,685,000,000
3,684,000,000
3,683,000,000
3,682,000,000
3,681,000,000

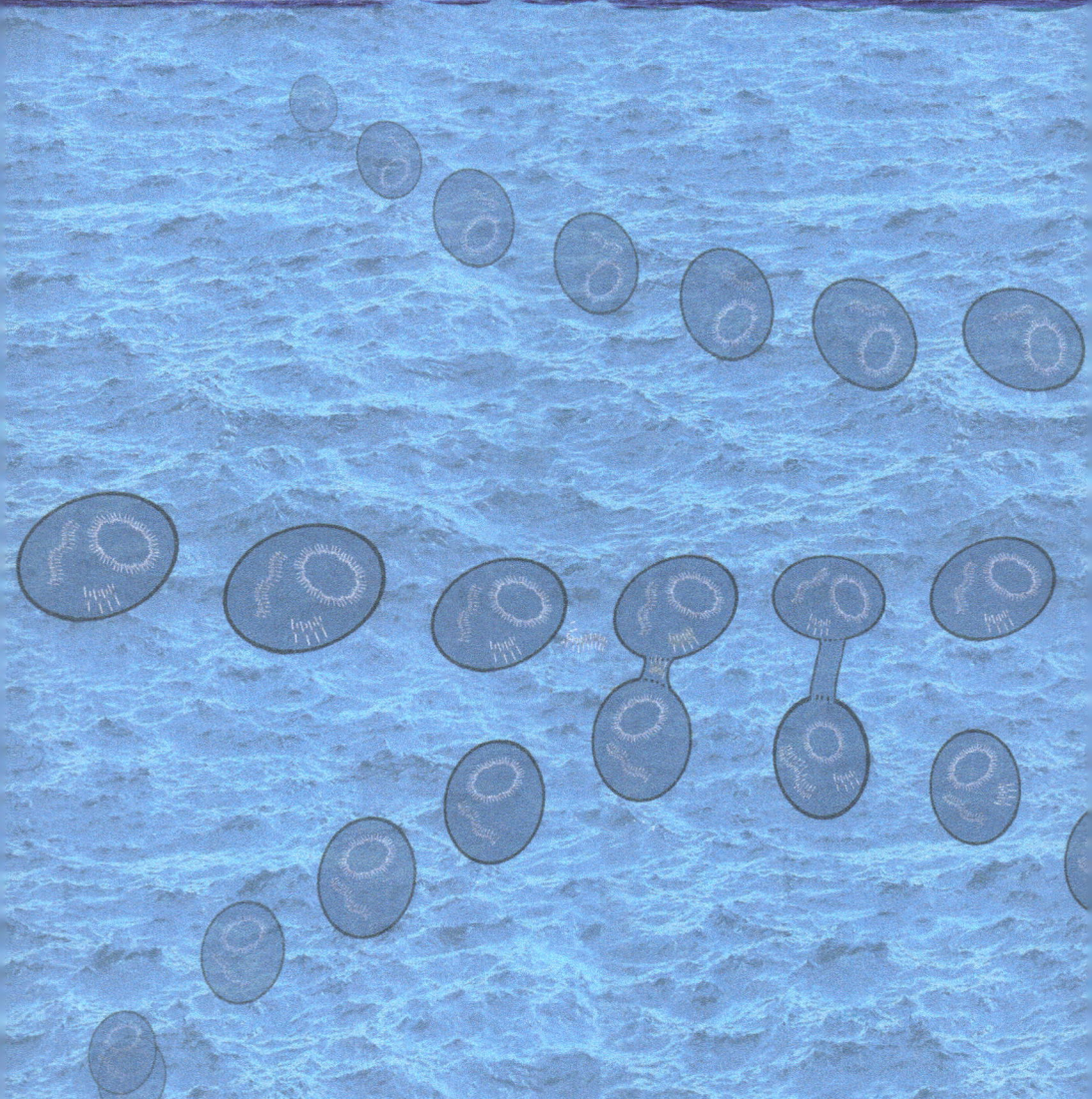

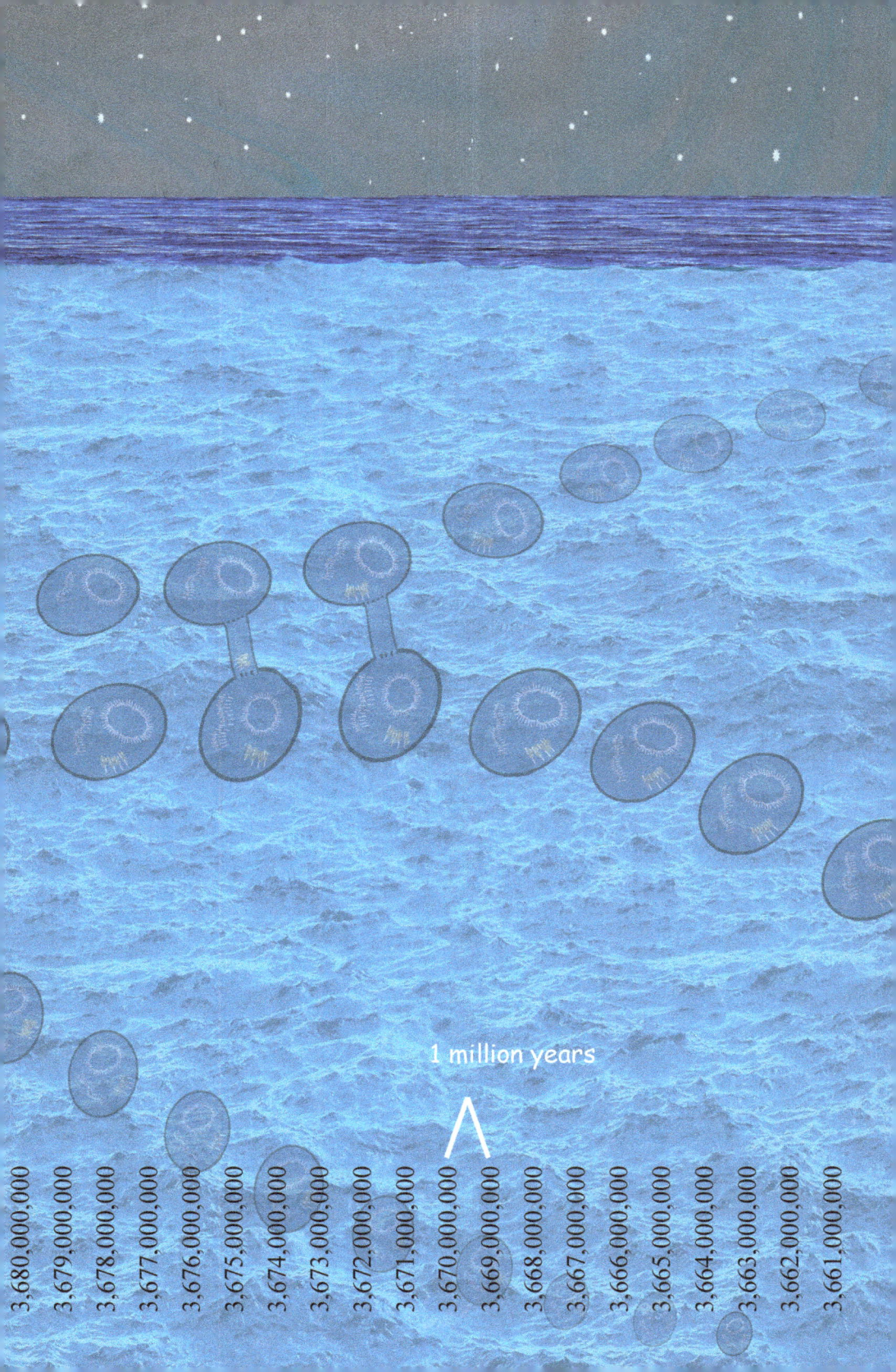

1 million years

3,680,000,000 3,679,000,000 3,678,000,000 3,677,000,000 3,676,000,000 3,675,000,000 3,674,000,000 3,673,000,000 3,672,000,000 3,671,000,000 3,670,000,000 3,669,000,000 3,668,000,000 3,667,000,000 3,666,000,000 3,665,000,000 3,664,000,000 3,663,000,000 3,662,000,000 3,661,000,000

We do a lot with nothing more than strands
of RNA. (We make proteins, we share
ourselves, metabolize and replicate.)
But in the course of time, mutations lead
nucleotides to recombine again,
each purine and pyrimidine attached
to its own link of an adjoining chain.

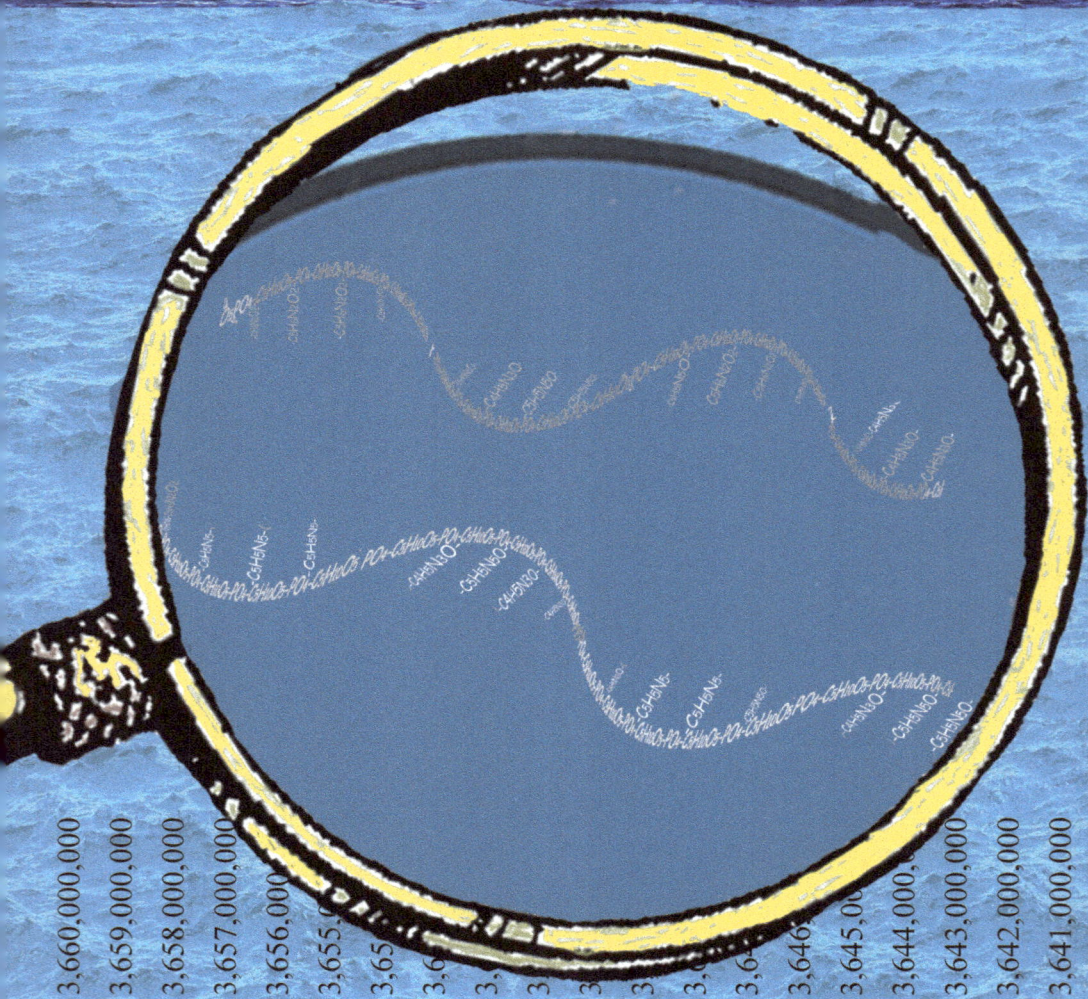

3,660,000,000
3,659,000,000
3,658,000,000
3,657,000,000
3,656,000,000
3,655,0
3,65
3,6
3,
3
3
3
3,646,
3,646
3,645,00
3,644,000,
3,643,000,000
3,642,000,000
3,641,000,000

These shapes twist into double helices
that someday will be known as DNA,
a stable molecule designed to last,
preserving the blueprints for who we are.

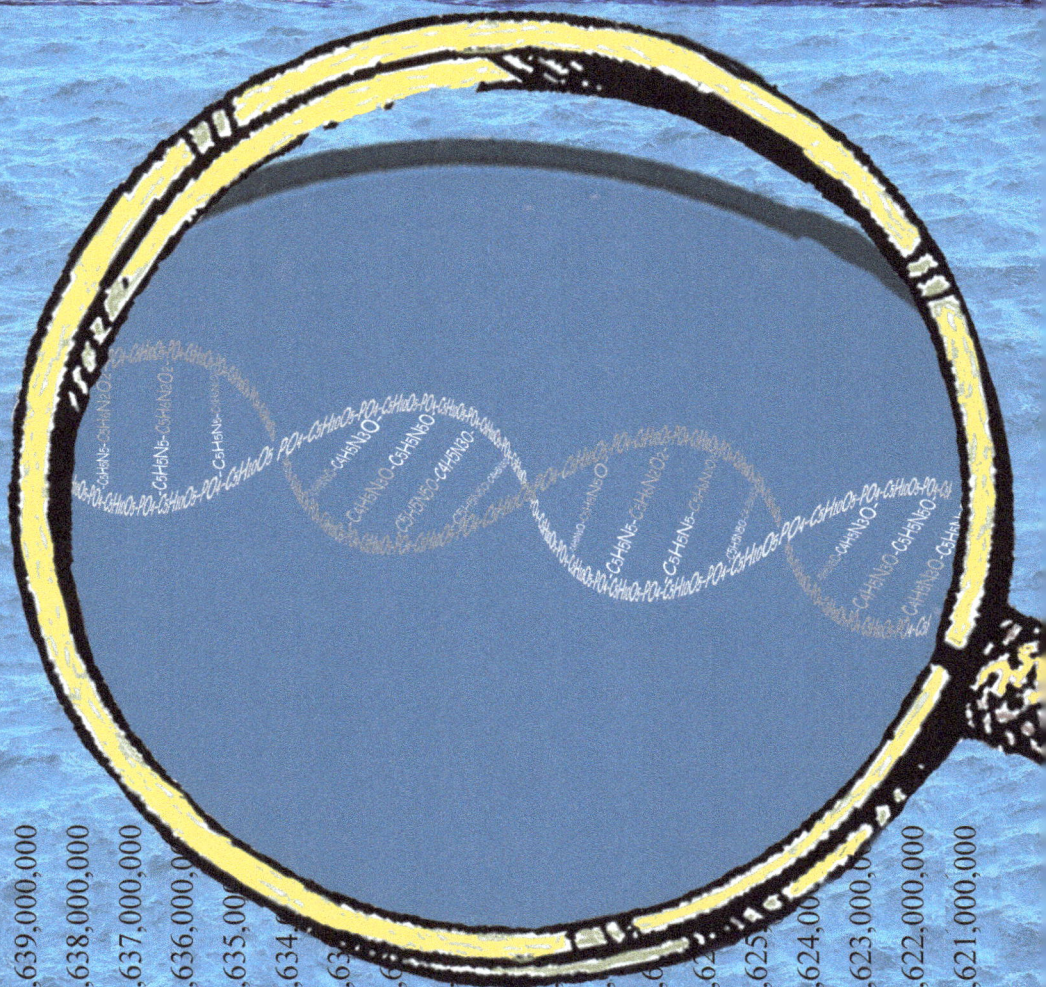

3,640,000,000
3,639,000,000
3,638,000,000
3,637,000,000
3,636,000,000
3,635,000,0
3,634
3,6
3,6
3,
3
3,
3,
3,
3,625
3,624,00
3,623,000,0
3,622,000,000
3,621,000,000

But even so, nucleic acids break
apart from time to time, and all the while,
the Sun emits harsh UV rays that block
our DNA; its photons interfere
with our sustained attempts to replicate.

3,620,
3,619,00
3,618,000,0
3,617,000,000
3,616,000,000
3,615,000,000
3,614,000,000
3,613,000,000
3,612,000,000
3,611,000,000
3,610,000,000
3,609,000,000
3,608,000,000
3,607,000,000
3,606,000,000
3,605,000,000
3,604,000,000
3,603,000,000
3,602,000,000
3,601,000,000

It takes a lot of luck — pure happenstance —
for mutant strands of DNA to be
a source of benefit within a cell,
so most of our mutations quickly fail.

3,600,000,000
3,599,000,000
3,598,000,000
3,597,000,000
3,596,000,000
3,595,000,000
3,594,000,000
3,593,000,000
3,592,000,000
3,591,000,000
3,590,000,000
3,589,000,000
3,588,000,000
3,587,000,000
3,586,000,000
3,585,000,000
3,584,000,000
3,583,000,000
3,582,000,000
3,581,000,000

But in the course of time, a few succeed,
and when they do, the newer DNA
persists and multiplies for centuries,
and — sharing these new genes — all of us gain.

3,580,000,000
3,579,000,000
3,578,000,000
3,577,000,000
3,576,000,000
3,575,000,000
3,574,000,000
3,573,000,000
3,572,000,000
3,571,000,000
3,570,000,000
3,569,000,000
3,568,000,000
3,567,000,000
3,566,000,000
3,565,000,000
3,564,000,000
3,563,000,000
3,562,000,000
3,561,000,000

Mutation means that most bacteria
and archaea have different chemistries,
but all of us remain prokaryotes
who share our DNA communally.
We all use energy from molecules
to make more lipid walls, more acid chains,
more stuff that makes our children like ourselves.

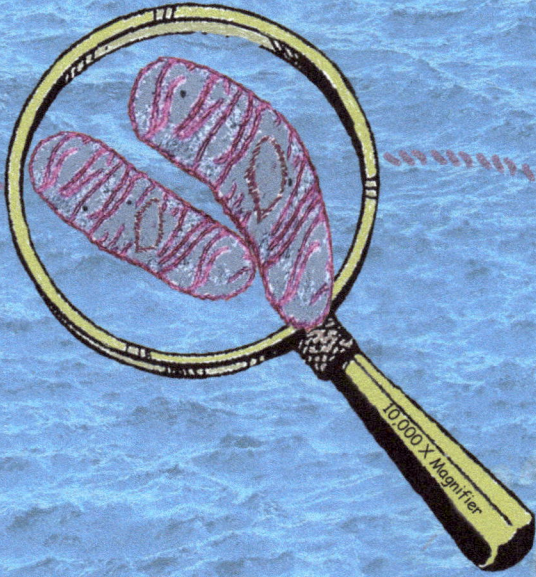

10,000 X Magnifier

3,540,000,000
3,539,000,000
3,538,000,000
3,537,000,000
3,536,000,000
3,535,000,000
3,534,000,000
3,533,000,000
3,532,000,000
3,531,000,000
3,530,000,000
3,529,000,000
3,528,000,000
3,527,000,000
3,526,000,000
3,525,000,000
3,524,000,000
3,523,000,000
3,522,000,000
3,521,000,000

And evolution takes a long, long time:
it's forty million years before it's time
to turn a page of life on Earth again.

3,520,000,000
3,519,000,000
3,518,000,000
3,517,000,000
3,516,000,000
3,515,000,000
3,514,000,000
3,513,000,000
3,512,000,000
3,511,000,000
3,510,000,000
3,509,000,000
3,508,000,000
3,507,000,000
3,506,000,000
3,505,000,000
3,504,000,000
3,503,000,000
3,502,000,000
3,501,000,000

For half a billion years we've fed ourselves (obtained our energy) from molecules alone. But now, five hundred million years since life on Earth began, a few of us devise a way to feed upon the Sun.

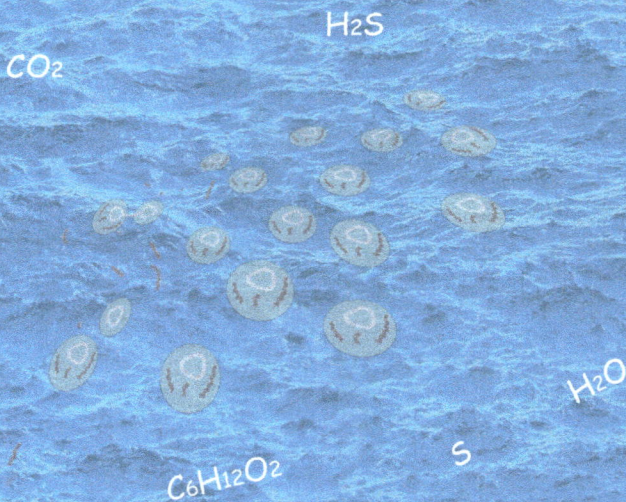

H_2S

CO_2

H_2O

S

$C_6H_{12}O_2$

$$H_2S + CO_2 \longrightarrow C_6H_{12}O_2 + H_2O + S$$

3,500,000,000
3,499,000,000
3,498,000,000
3,497,000,000
3,496,000,000
3,495,000,000
3,494,000,000
3,493,000,000
3,492,000,000
3,491,000,000
3,490,000,000
3,489,000,000
3,488,000,000
3,487,000,000
3,486,000,000
3,485,000,000
3,484,000,000
3,483,000,000
3,482,000,000
3,481,000,000

H₂S

CO₂

H₂O

C₆H₁₂O₆

S

$$H_2S + CO_2 + Light \longrightarrow C_6H_{12}O_6 + H_2O + S$$

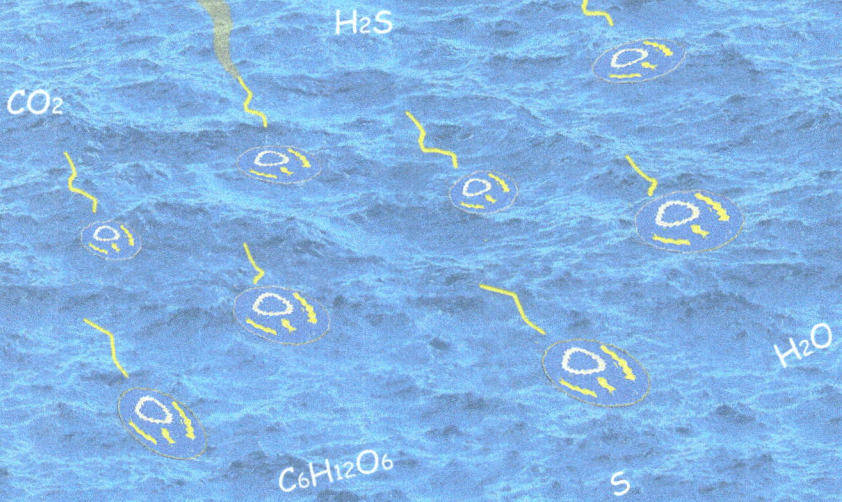

3,480,000,000
3,479,000,000
3,478,000,000
3,477,000,000
3,476,000,000
3,475,000,000
3,474,000,000
3,473,000,000
3,472,000,000
3,471,000,000
3,470,000,000
3,469,000,000
3,468,000,000
3,467,000,000
3,466,000,000
3,465,000,000
3,464,000,000
3,463,000,000
3,462,000,000
3,461,000,000

3,460,000,000
3,459,000,000
3,458,000,000
3,457,000,000
3,456,000,000
3,455,000,000
3,454,000,000
3,453,000,000
3,452,000,000
3,451,000,000
3,450,000,000
3,449,000,000
3,448,000,000
3,447,000,000
3,446,000,000
3,445,000,000
3,444,000,000
3,443,000,000
3,442,000,000
3,441,000,000

"Anoxygenic photosynthesis"
involves a very different way to live.
It doesn't produce deadly oxygen
but it provides more needed energy
than molecules alone have ever done.

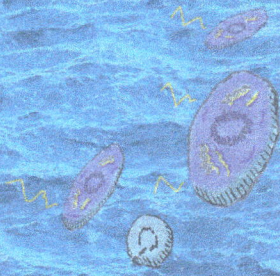

First anoxygenic photosynthesis about 3.4 bya

3,440,000,000
3,439,000,000
3,438,000,000
3,437,000,000
3,436,000,000
3,435,000,000
3,434,000,000
3,433,000,000
3,432,000,000
3,431,000,000
3,430,000,000
3,429,000,000
3,428,000,000
3,427,000,000
3,426,000,000
3,425,000,000
3,424,000,000
3,423,000,000
3,422,000,000
3,421,000,000

And as each million years goes speeding by,
we replicate ourselves as best we can.

3,420,000,000
3,419,000,000
3,418,000,000
3,417,000,000
3,416,000,000
3,415,000,000
3,414,000,000
3,413,000,000
3,412,000,000
3,411,000,000
3,410,000,000
3,409,000,000
3,408,000,000
3,407,000,000
3,406,000,000
3,405,000,000
3,404,000,000
3,403,000,000
3,402,000,000
3,401,000,000

We do not swim, because we have no fins
nor any wiggling parts with which to move.
We're subject to the ebb and flow of tides,
transported by the movements of the sea.

3,400,000,000
3,399,000,000
3,398,000,000
3,397,000,000
3,396,000,000
3,395,000,000
3,394,000,000
3,393,000,000
3,392,000,000
3,391,000,000
3,390,000,000
3,389,000,000
3,388,000,000
3,387,000,000
3,386,000,000
3,385,000,000
3,384,000,000
3,383,000,000
3,382,000,000
3,381,000,000

While some of us sink down to where it's dark and everything is cold, eternal night,

3,380,000,000
3,379,000,000
3,378,000,000
3,377,000,000
3,376,000,000
3,375,000,000
3,374,000,000
3,373,000,000
3,372,000,000
3,371,000,000
3,370,000,000
3,369,000,000
3,368,000,000
3,367,000,000
3,366,000,000
3,365,000,000
3,364,000,000
3,363,000,000
3,362,000,000
3,361,000,000

and those of us with lighter body weights rise closer to the surface and the light, mutation tends to give us what we need to thrive in different habitats.

3,360,000,000
3,359,000,000
3,358,000,000
3,357,000,000
3,356,000,000
3,355,000,000
3,354,000,000
3,353,000,000
3,352,000,000
3,351,000,000
3,350,000,000
3,349,000,000
3,348,000,000
3,347,000,000
3,346,000,000
3,345,000,000
3,344,000,000
3,343,000,000
3,342,000,000
3,341,000,000

For some
it's still those steamy hydrothermal vents,
while others make new chemistries in ice.
A few adapt to salt; others acidity;
still others find that sulfur has appeal,
while some prefer to live on methane gas.

H_2S HSO_4

3,340,000,000
3,339,000,000
3,338,000,000
3,337,000,000
3,336,000,000
3,335,000,000
3,334,000,000
3,333,000,000
3,332,000,000
3,331,000,000
3,330,000,000
3,329,000,000
3,328,000,000
3,327,000,000
3,326,000,000
3,325,000,000
3,324,000,000
3,323,000,000
3,322,000,000
3,321,000,000

CH₄

CH₄

3,320,000,000
3,319,000,000
3,318,000,000
3,317,000,000
3,316,000,000
3,315,000,000
3,314,000,000
3,313,000,000
3,312,000,000
3,311,000,000
3,310,000,000
3,309,000,000
3,308,000,000
3,307,000,000
3,306,000,000
3,305,000,000
3,304,000,000
3,303,000,000
3,302,000,000
3,301,000,000

Not one of us requires free oxygen (there's none of it around) and each of us gets by just fine without the deadly stuff. And all the while, mutation lets us all adapt to the great challenges of life by changing our ideas of how to live.

3.300.000.000
3.299.000.000
3.298.000.000
3.297.000.000
3.296.000.000
3.295.000.000
3.294.000.000
3.293.000.000
3.292.000.000
3.291.000.000
3.290.000.000
3.289.000.000
3.288.000.000
3.287.000.000
3.286.000.000
3.285.000.000
3.284.000.000
3.283.000.000
3.282.000.000
3.281.000.000

3,280,000,000
3,279,000,000
3,278,000,000
3,277,000,000
3,276,000,000
3,275,000,000
3,274,000,000
3,273,000,000
3,272,000,000
3,271,000,000
3,270,000,000
3,269,000,000
3,268,000,000
3,267,000,000
3,266,000,000
3,265,000,000
3,264,000,000
3,263,000,000
3,262,000,000
3,261,000,000

3. EXPANSION

Now, deep below the surface of the sea,
Earth's mantle and its crust still shift and grate
and push the ocean floor up through the waves.
And in the course of time, the hills that form
above grow high enough that islands join
together, water drains, and land begins
to take up ever more of what's above,
until it forms into a vast dry mass
that we will someday label "Vaalbara."

3,260,000,000
3,259,000,000
3,258,000,000
3,257,000,000
3,256,000,000
3,255,000,000
3,254,000,000
3,253,000,000
3,252,000,000
3,251,000,000
3,250,000,000
3,249,000,000
3,248,000,000
3,247,000,000
3,246,000,000
3,245,000,000
3,244,000,000
3,243,000,000
3,242,000,000
3,241,000,000

CO₂

CO₂

While CO_2 (the stuff we love) abounds
above the surface of the sea, the air
up there is thin and dry and Vaalbara
strikes none of us as any place to live.

3,240,000,000
3,239,000,000
3,238,000,000
3,237,000,000
3,236,000,000
3,235,000,000
3,234,000,000
3,233,000,000
3,232,000,000
3,231,000,000
3,230,000,000
3,229,000,000
3,228,000,000
3,227,000,000
3,226,000,000
3,225,000,000
3,224,000,000
3,223,000,000
3,222,000,000
3,221,000,000

For all of us remain small, simple things,
just RNA and DNA, encased
in lipid walls — not mariners with oars,
nor paddlers in canoes, not even blobs
of iellyfish — just single cells without
the means to wiggle, navigate or climb.

3,220,000,000
3,219,000,000
3,218,000,000
3,217,000,000
3,216,000,000
3,215,000,000
3,214,000,000
3,213,000,000
3,212,000,000
3,211,000,000
3,210,000,000
3,209,000,000
3,208,000,000
3,207,000,000
3,206,000,000
3,205,000,000
3,204,000,000
3,203,000,000
3,202,000,000
3,201,000,000

Like flotsam: on our own and very small.

3,200,000,000
3,199,000,000
3,198,000,000
3,197,000,000
3,196,000,000
3,195,000,000
3,194,000,000
3,193,000,000
3,192,000,000
3,191,000,000
3,190,000,000
3,189,000,000
3,188,000,000
3,187,000,000
3,186,000,000
3,185,000,000
3,184,000,000
3,183,000,000
3,182,000,000
3,181,000,000

The winds and tides and currents have their way:
they rock us gently with their every move.
Our lipid walls aren't made to keep us safe
from air, or land, or anything but sea.

3,180,000,000
3,179,000,000
3,178,000,000
3,177,000,000
3,176,000,000
3,175,000,000
3,174,000,000
3,173,000,000
3,172,000,000
3,171,000,000
3,170,000,000
3,169,000,000
3,168,000,000
3,167,000,000
3,166,000,000
3,165,000,000
3,164,000,000
3,163,000,000
3,162,000,000
3,161,000,000

When some of us get tossed up onto land, exposure to the air means we dry out: we lose our vital liquids, and we die upon the brutal shores of Vaalbara.

3,160,000,000
3,159,000,000
3,158,000,000
3,157,000,000
3,156,000,000
3,155,000,000
3,154,000,000
3,153,000,000
3,152,000,000
3,151,000,000
3,150,000,000
3,149,000,000
3,148,000,000
3,147,000,000
3,146,000,000
3,145,000,000
3,144,000,000
3,143,000,000
3,142,000,000
3,141,000,000

3,560,000,000 · 3,559,000,000 · 3,558,000,000 · 3,557,000,000 · 3,556,000,000 · 3,555,000,000 · 3,554,000,000 · 3,553,000,000 · 3,552,000,000 · 3,551,000,000 · 3,550,000,000 · 3,549,000,000 · 3,548,000,000 · 3,547,000,000 · 3,546,000,000 · 3,545,000,000 · 3,544,000,000 · 3,543,000,000 · 3,542,000,000 · 3,541,000,000

It seems it's always safer to remain
where everything is as it's always been.
Prokaryotic life has always done
alright in this vast, anaerobic sea
where we were born. As long as we've stayed put,
there's been no catastrophic end to things.

3,140,000,000
3,139,000,000
3,138,000,000
3,137,000,000
3,136,000,000
3,135,000,000
3,134,000,000
3,133,000,000
3,132,000,000
3,131,000,000
3,130,000,000
3,129,000,000
3,128,000,000
3,127,000,000
3,126,000,000
3,125,000,000
3,124,000,000
3,123,000,000
3,122,000,000
3,121,000,000

For all it takes for chemicals to thrive is chemistry, and chemicals don't go extinct. Our kind — prokaryotes — remain alive and doing rather well as we approach our billionth year on Earth.

3,120,000,000
3,119,000,000
3,118,000,000
3,117,000,000
3,116,000,000
3,115,000,000
3,114,000,000
3,113,000,000
3,112,000,000
3,111,000,000
3,110,000,000
3,109,000,000
3,108,000,000
3,107,000,000
3,106,000,000
3,105,000,000
3,104,000,000
3,103,000,000
3,102,000,000
3,101,000,000

But in the randomness of chains that break
and join again, there comes a time, against
all odds, when some of us form stronger walls
that keep our vital liquids safe within —
so when we're next cast up on arid shores,
we manage to survive the dreadful place.

3,100,000,000
3,099,000,000
3,098,000,000
3,097,000,000
3,096,000,000
3,095,000,000
3,094,000,000
3,093,000,000
3,092,000,000
3,091,000,000
3,090,000,000
3,089,000,000
3,088,000,000
3,087,000,000
3,086,000,000
3,085,000,000
3,084,000,000
3,083,000,000
3,082,000,000
3,081,000,000

3,080,000,000
3,079,000,000
3,078,000,000
3,077,000,000
3,076,000,000
3,075,000,000
3,074,000,000
3,073,000,000
3,072,000,000
3,071,000,000
3,070,000,000
3,069,000,000
3,068,000,000
3,067,000,000
3,066,000,000
3,065,000,000
3,064,000,000
3,063,000,000
3,062,000,000
3,061,000,000

And carried on the wind by gusts of air
that, light and dry, are rife with CO_2,
we're strewn from sunny peak to canyon floor.

3,060,000,000
3,059,000,000
3,058,000,000
3,057,000,000
3,056,000,000
3,055,000,000
3,054,000,000
3,053,000,000
3,052,000,000
3,051,000,000
3,050,000,000
3,049,000,000
3,048,000,000
3,047,000,000
3,046,000,000
3,045,000,000
3,044,000,000
3,043,000,000
3,042,000,000
3,041,000,000

So don't mistake us for amphibians
or worms with legs — we're still just tiny cells,
the slightest film of microbes on the rocks.

3,040,000,000
3,039,000,000
3,038,000,000
3,037,000,000
3,036,000,000
3,035,000,000
3,034,000,000
3,033,000,000
3,032,000,000
3,031,000,000
3,030,000,000
3,029,000,000
3,028,000,000
3,027,000,000
3,026,000,000
3,025,000,000
3,024,000,000
3,023,000,000
3,022,000,000
3,021,000,000

3,020,000,000 3,019,000,000 3,018,000,000 3,017,000,000 3,016,000,000 3,015,000,000 3,014,000,000 3,013,000,000 3,012,000,000 3,011,000,000 3,010,000,000 3,009,000,000 3,008,000,000 3,007,000,000 3,006,000,000 3,005,000,000 3,004,000,000 3,003,000,000 3,002,000,000 3,001,000,000

And as we propagate on Earth's dry stones,
we learn about the splendor of the rain,
whose drops make rivers, lakes, and inland bays
and make us feel as if we'd never left
those first warm seas in which our lives began.

3,000,000,000
2,999,000,000
2,998,000,000
2,997,000,000
2,996,000,000
2,995,000,000
2,994,000,000
2,993,000,000
2,992,000,000
2,991,000,000
2,990,000,000
2,989,000,000
2,988,000,000
2,987,000,000
2,986,000,000
2,985,000,000
2,984,000,000
2,983,000,000
2,982,000,000
2,981,000,000

We replicate perhaps a billion times
for every million years that we exist.

2,980,000,000
2,979,000,000
2,978,000,000
2,977,000,000
2,976,000,000
2,975,000,000
2,974,000,000
2,973,000,000
2,972,000,000
2,971,000,000
2,970,000,000
2,969,000,000
2,968,000,000
2,967,000,000
2,966,000,000
2,965,000,000
2,964,000,000
2,963,000,000
2,962,000,000
2,961,000,000

1,000,000,000 replications each
∧

2,960,000,000
2,959,000,000
2,958,000,000
2,957,000,000
2,956,000,000
2,955,000,000
2,954,000,000
2,953,000,000
2,952,000,000
2,951,000,000
2,950,000,000
2,949,000,000
2,948,000,000
2,947,000,000
2,946,000,000
2,945,000,000
2,944,000,000
2,943,000,000
2,942,000,000
2,941,000,000

2,940,000,000
2,939,000,000
2,938,000,000
2,937,000,000
2,936,000,000
2,935,000,000
2,934,000,000
2,933,000,000
2,932,000,000
2,931,000,000
2,930,000,000
2,929,000,000
2,928,000,000
2,927,000,000
2,926,000,000
2,925,000,000
2,924,000,000
2,923,000,000
2,922,000,000
2,921,000,000

2,920,000,000
2,919,000,000
2,918,000,000
2,917,000,000
2,916,000,000
2,915,000,000
2,914,000,000
2,913,000,000
2,912,000,000
2,911,000,000
2,910,000,000
2,909,000,000
2,908,000,000
2,907,000,000
2,906,000,000
2,905,000,000
2,904,000,000
2,903,000,000
2,902,000,000
2,901,000,000

Our chemistries adapt to heat, to cold,
to light and dark, but always, we retain
our structure: molecules of RNA
and DNA enclosed in lipid walls,
afloat within a cytoplasmic sea
about a single, tiny micron long.

2,900,000,000
2,899,000,000
2,898,000,000
2,897,000,000
2,896,000,000
2,895,000,000
2,894,000,000
2,893,000,000
2,892,000,000
2,891,000,000
2,890,000,000
2,889,000,000
2,888,000,000
2,887,000,000
2,886,000,000
2,885,000,000
2,884,000,000
2,883,000,000
2,882,000,000
2,881,000,000

2,880,000,000
2,879,000,000
2,878,000,000
2,877,000,000
2,876,000,000
2,875,000,000
2,874,000,000
2,873,000,000
2,872,000,000
2,871,000,000
2,870,000,000
2,869,000,000
2,868,000,000
2,867,000,000
2,866,000,000
2,865,000,000
2,864,000,000
2,863,000,000
2,862,000,000
2,861,000,000

No arms or legs, no hearts or brains or mouths.

2,860,000,000
2,859,000,000
2,858,000,000
2,857,000,000
2,856,000,000
2,855,000,000
2,854,000,000
2,853,000,000
2,852,000,000
2,851,000,000
2,850,000,000
2,849,000,000
2,848,000,000
2,847,000,000
2,846,000,000
2,845,000,000
2,844,000,000
2,843,000,000
2,842,000,000
2,841,000,000

That's what it means to be prokaryotes.

2,840,000,000
2,839,000,000
2,838,000,000
2,837,000,000
2,836,000,000
2,835,000,000
2,834,000,000
2,833,000,000
2,832,000,000
2,831,000,000
2,830,000,000
2,829,000,000
2,828,000,000
2,827,000,000
2,826,000,000
2,825,000,000
2,824,000,000
2,823,000,000
2,822,000,000
2,821,000,000

CO₂

H₂S

H₂S + CO₂ → C₆H₁₂O₂ + H₂O + S

H₂S

CO₂

2,820,000,000 2,819,000,000 2,818,000,000 2,817,000,000 2,816,000,000 2,815,000,000 2,814,000,000 2,813,000,000 2,812,000,000 2,811,000,000 2,810,000,000 2,809,000,000 2,808,000,000 2,807,000,000 2,806,000,000 2,805,000,000 2,804,000,000 2,803,000,000 2,802,000,000 2,801,000,000

We're only chemicals within a shell, exchanging electrons and making new metabolites from all that CO_2 and hydrogen, to get the energy to replicate ourselves.

H_2O

H_2O

S

S

2,800,000,000
2,799,000,000
2,798,000,000
2,797,000,000
2,796,000,000
2,795,000,000
2,794,000,000
2,793,000,000
2,792,000,000
2,791,000,000
2,790,000,000
2,789,000,000
2,788,000,000
2,787,000,000
2,786,000,000
2,785,000,000
2,784,000,000
2,783,000,000
2,782,000,000
2,781,000,000

4. OXYGENESIS

Now, as years pass
in our microbial world, the period
they'll someday call "Methanian" arrives,
the end result of many years in which
methanogens have steadily increased
the methane here on Earth.

CH₄

CH₄

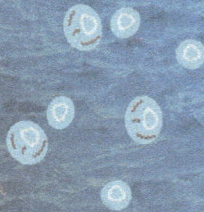

CH₄

2,780,000,000
2,779,000,000
2,778,000,000
2,777,000,000
2,776,000,000
2,775,000,000
2,774,000,000
2,773,000,000
2,772,000,000
2,771,000,000
2,770,000,000
2,769,000,000
2,768,000,000
2,767,000,000
2,766,000,000
2,765,000,000
2,764,000,000
2,763,000,000
2,762,000,000
2,761,000,000

But we've had time
(a hundred million years or so) to get
accustomed to this change, so life goes on.
By sharing DNA, and staying small,
we can adapt, for chemistry evolves
one molecule, one micron at a time.

CH4

CH4

CH4

CH4

2,740,000,000
2,739,000,000
2,738,000,000
2,737,000,000
2,736,000,000
2,735,000,000
2,734,000,000
2,733,000,000
2,732,000,000
2,731,000,000
2,730,000,000
2,729,000,000
2,728,000,000
2,727,000,000
2,726,000,000
2,725,000,000
2,724,000,000
2,723,000,000
2,722,000,000
2,721,000,000

Now, we've spent thirteen hundred million years
immersed in H_2O and CO_2
(and lately methane gas) without a need
to breathe free oxygen, when some of us —
the cyanobacteria — look once
again to sunlight as a source of power.

2,720,000,000
2,719,000,000
2,718,000,000
2,717,000,000
2,716,000,000
2,715,000,000
2,714,000,000
2,713,000,000
2,712,000,000
2,711,000,000
2,710,000,000
2,709,000,000
2,708,000,000
2,707,000,000
2,706,000,000
2,705,000,000
2,704,000,000
2,703,000,000
2,702,000,000
2,701,000,000

But photosynthesis is different now,
done cyanobacteria's new way.

First oxygenic photosynthesis
about 2.7 bya

2,700,000,000
2,699,000,000
2,698,000,000
2,697,000,000
2,696,000,000
2,695,000,000
2,694,000,000
2,693,000,000
2,692,000,000
2,691,000,000
2,690,000,000
2,689,000,000
2,688,000,000
2,687,000,000
2,686,000,000
2,685,000,000
2,684,000,000
2,683,000,000
2,682,000,000
2,681,000,000

2,680,000,000
2,679,000,000
2,678,000,000
2,677,000,000
2,676,000,000
2,675,000,000
2,674,000,000
2,673,000,000
2,672,000,000
2,671,000,000
2,670,000,000
2,669,000,000
2,668,000,000
2,667,000,000
2,666,000,000
2,665,000,000
2,664,000,000
2,663,000,000
2,662,000,000
2,661,000,000

These microbes start with simple H_2O
and CO_2. When energized by light,
they turn them into sugar, which is food.
The process yields a lot of energy,
but there's an unintended consequence,
a useless byproduct: free oxygen.

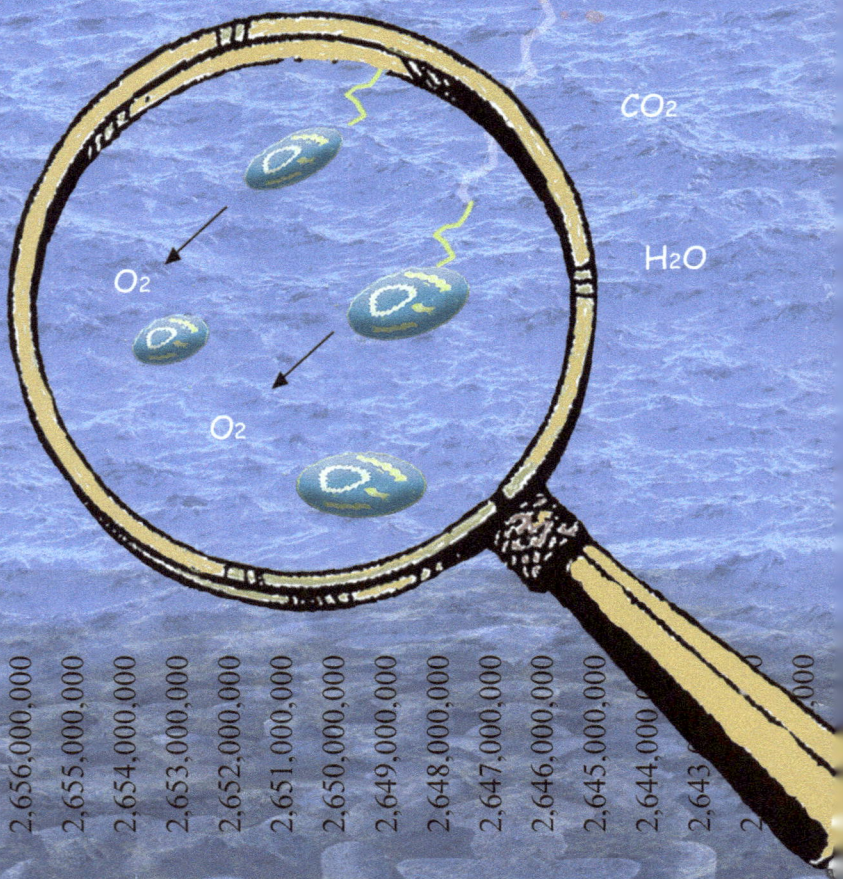

CO_2

H_2O

O_2

O_2

2,660,000,000
2,659,000,000
2,658,000,000
2,657,000,000
2,656,000,000
2,655,000,000
2,654,000,000
2,653,000,000
2,652,000,000
2,651,000,000
2,650,000,000
2,649,000,000
2,648,000,000
2,647,000,000
2,646,000,000
2,645,000,000
2,644,000,
2,643

At first, the noxious waste is hardly felt;
the oxygen combines with hydrogen
to make more H_2O.

H_2

H_2

O_2

O_2

$O + H_2 = H_2O$

$O + H_2 = H_2O$

2,640,000,000
2,639,000,000
2,638,000,000
2,637,000,000
2,636,000,000
2,635,000,000
2,634,000,000
2,633,000,000
2,632,000,000
2,631,000,000
2,630,000,000
2,629,000,000
2,628,000,000
2,627,000,000
2,626,000,000
2,625,000,000
2,624,000,000
2,623,000,000
2,622,000,000
2,621,000,000

But in the course
of time — a hundred million years or so —
this drawing on the power of the sun
provides such boundless energy it proves
a boon to those who photosynthesize,
and ever more of us use solar power
to drive the workings of our lives.

2,620,000,000
2,619,000,000
2,618,000,000
2,617,000,000
2,616,000,000
2,615,000,000
2,614,000,000
2,613,000,000
2,612,000,000
2,611,000,000
2,610,000,000
2,609,000,000
2,608,000,000
2,607,000,000
2,606,000,000
2,605,000,000
2,604,000,000
2,603,000,000
2,602,000,000
2,601,000,000

2,600,000,000
2,599,000,000
2,598,000,000
2,597,000,000
2,596,000,000
2,595,000,000
2,594,000,000
2,593,000,000
2,592,000,000
2,591,000,000
2,590,000,000
2,589,000,000
2,588,000,000
2,587,000,000
2,586,000,000
2,585,000,000
2,584,000,000
2,583,000,000
2,582,000,000
2,581,000,000

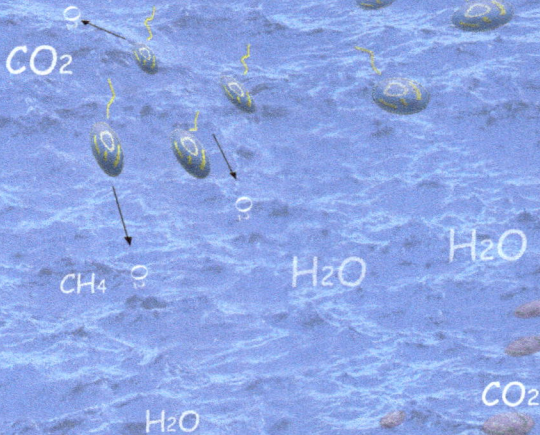

There comes
a time when all this oxygen exceeds
what molecules like hydrogen absorb,
and as it fills the seas, it threatens all
of us who've spent more than a billion years
adapting to the anaerobic world.

CO_2

CH_4

H_2O

H_2O

H_2O

CO_2

CO_2

2,580,000,000
2,579,000,000
2,578,000,000
2,577,000,000
2,576,000,000
2,575,000,000
2,574,000,000
2,573,000,000
2,572,000,000
2,571,000,000
2,570,000,000
2,569,000,000
2,568,000,000
2,567,000,000
2,566,000,000
2,565,000,000
2,564,000,000
2,563,000,000
2,562,000,000
2,561,000,000

O₂

O₂

O₂

O₂

O₂

O₂

2,560,000,000
2,559,000,000
2,558,000,000
2,557,000,000
2,556,000,000
2,555,000,000
2,554,000,000
2,553,000,000
2,552,000,000
2,551,000,000
2,550,000,000
2,549,000,000
2,548,000,000
2,547,000,000
2,546,000,000
2,545,000,000
2,544,000,000
2,543,000,000
2,542,000,000
2,541,000,000

We living things aren't used to living in the midst of toxic waste. But now, the time has come: we *must* adapt, and so we do.

O_2

H_2O

CH_4

H_2O

CO_2

2,540,000,000
2,539,000,000
2,538,000,000
2,537,000,000
2,536,000,000
2,535,000,000
2,534,000,000
2,533,000,000
2,532,000,000
2,531,000,000
2,530,000,000
2,529,000,000
2,528,000,000
2,527,000,000
2,526,000,000
2,525,000,000
2,524,000,000
2,523,000,000
2,522,000,000
2,521,000,000

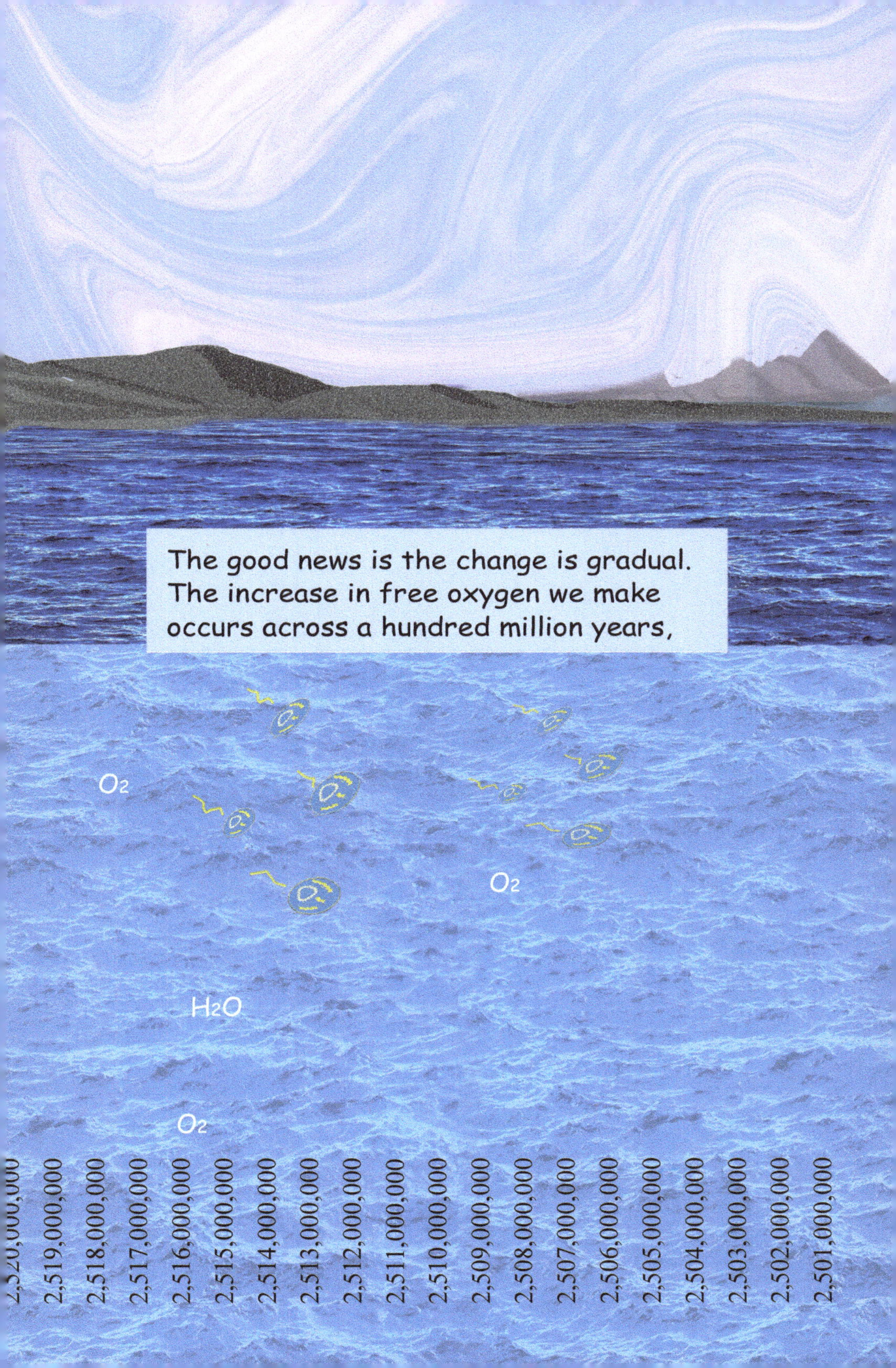

The good news is the change is gradual.
The increase in free oxygen we make
occurs across a hundred million years,
O2
O2
H2O
O2
2,519,000,000
2,518,000,000
2,517,000,000
2,516,000,000
2,515,000,000
2,514,000,000
2,513,000,000
2,512,000,000
2,511,000,000
2,510,000,000
2,509,000,000
2,508,000,000
2,507,000,000
2,506,000,000
2,505,000,000
2,504,000,000
2,503,000,000
2,502,000,000
2,501,000,000

which gives us time to try new sequences of DNA to cope with all the waste.

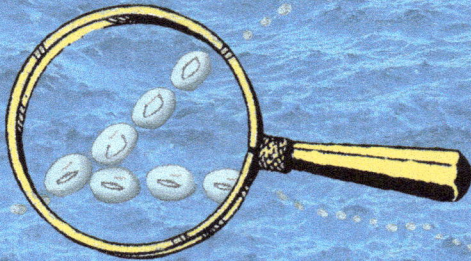

2,500,000,000
2,499,000,000
2,498,000,000
2,497,000,000
2,496,000,000
2,495,000,000
2,494,000,000
2,493,000,000
2,492,000,000
2,491,000,000
2,490,000,000
2,489,000,000
2,488,000,000
2,487,000,000
2,486,000,000
2,485,000,000
2,484,000,000
2,483,000,000
2,482,000,000
2,481,000,000

Some can't survive the metamorphosis, but some adapt, and benefit.

2,460,000,000
2,459,000,000
2,458,000,000
2,457,000,000
2,456,000,000
2,455,000,000
2,454,000,000
2,453,000,000
2,452,000,000
2,451,000,000
2,450,000,000
2,449,000,000
2,448,000,000
2,447,000,000
2,446,000,000
2,445,000,000
2,444,000,000
2,443,000,000
2,442,000,000
2,441,000,000

2,440,000,000
2,439,000,000
2,438,000,000
2,437,000,000
2,436,000,000
2,435,000,000
2,434,000,000
2,433,000,000
2,432,000,000
2,431,000,000
2,430,000,000
2,429,000,000
2,428,000,000
2,427,000,000
2,426,000,000
2,425,000,000
2,424,000,000
2,423,000,000
2,422,000,000
2,421,000,000

In time,
while changes in the world in which we live
bring glaciers, sometimes even "snowball earths,"
the very rock that forms the ocean floor
absorbs its share of excess oxygen.

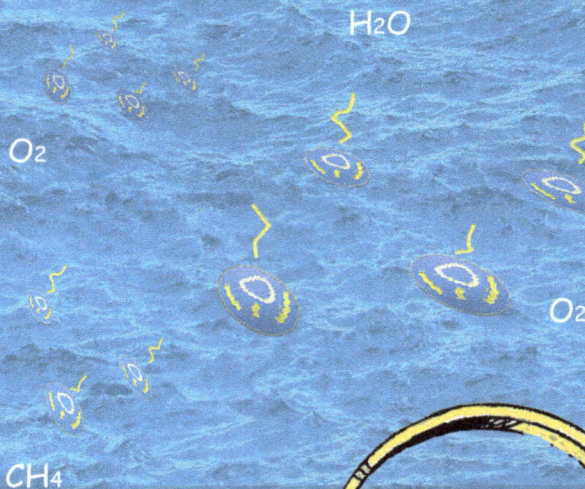

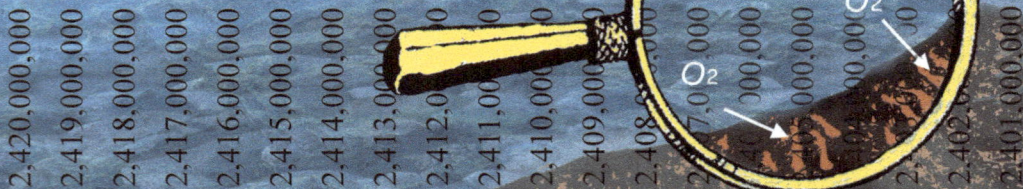

H_2O

O_2

O_2

CH_4

O_2

O_2

2,420,000,000
2,419,000,000
2,418,000,000
2,417,000,000
2,416,000,000
2,415,000,000
2,414,000,000
2,413,000,000
2,412,000,000
2,411,000,000
2,410,000,000
2,409,000,000
2,408,000,000
2,407,000,000
2,000,000
2,000,000
2,402,000,000
2,401,000,000

O_2
O_2
O_2
O_2
O_2
O_2
O_2
O_2
O_2
O_2
O_2
O_2
O_2
O_2
O_2
O_2
O_2
O_2
O_2
O_2
O_2
O_2

Still, photosynthesis goes on. So when
the hydrogen and ocean floor have done
as much as they can to absorb their share
of all that noxious waste, the oxygen
begins to make its way into the air.

O_2

O_2

O_2

O_2

First evidence of free oxygen
in the atmosphere 2.4 bya

2,400,000,000
2,399,000,000
2,398,000,000
2,397,000,000
2,396,000,000
2,395,000,000
2,394,000,000
2,393,000,000
2,392,000,000
2,391,000,000
2,390,000,000
2,389,000,000
2,388,000,000
2,387,000,000
2,386,000,000
2,385,000,000
2,384,000,000
2,383,000,000
2,382,000,000
2,381,000,000

We cyanobacteria are still
just single cells, about a micron long,
but photosynthesis so benefits
our kind that soon we've spread around the globe.

2,380,000,000
2,379,000,000
2,378,000,000
2,377,000,000
2,376,000,000
2,375,000,000
2,374,000,000
2,373,000,000
2,372,000,000
2,371,000,000
2,370,000,000
2,369,000,000
2,368,000,000
2,367,000,000
2,366,000,000
2,365,000,000
2,364,000,000
2,363,000,000
2,362,000,000
2,361,000,000

O₂ O₂ O₂ O₂ O₂ O₂ O₂ O₂ O₂ O₂ O₂ O₂ O₂ O₂ O₂ O₂

O_2 O_2 O_2 O_2 O_2 O_2 O_2 O_2 O_2 O_2 O_2 O_2 O_2 O_2 O_2 O_2

2,360,000,000
2,359,000,000
2,358,000,000
2,357,000,000
2,356,000,000
2,355,000,000
2,354,000,000
2,353,000,000
2,352,000,000
2,351,000,000
2,350,000,000
2,349,000,000
2,348,000,000
2,347,000,000
2,346,000,000
2,345,000,000
2,344,000,000
2,343,000,000
2,342,000,000
2,341,000,000

O_2 O_2 O_2
O_2 O_2 O_2
O_2 O_2 O_2
O_2 O_2 O_2
O_2 O_2 O_2
O_2 O_2 O_2
O_2 O_2 O_2
O_2 O_2
O_2
O_2
O_2

Vast mats of microbes lurk beneath the waves,

2,340,000,000
2,339,000,000
2,338,000,000
2,337,000,000
2,336,000,000
2,335,000,000
2,334,000,000
2,333,000,000
2,332,000,000
2,331,000,000
2,330,000,000
2,329,000,000
2,328,000,000
2,327,000,000
2,326,000,000
2,325,000,000
2,324,000,000
2,323,000,000
2,322,000,000
2,321,000,000

So many billions of us, far and wide,
that one can likely see us from the moon.

2,320,000,000
2,319,000,000
2,318,000,000
2,317,000,000
2,316,000,000
2,315,000,000
2,314,000,000
2,313,000,000
2,312,000,000
2,311,000,000
2,310,000,000
2,309,000,000
2,308,000,000
2,307,000,000
2,306,000,000
2,305,000,000
2,304,000,000
2,303,000,000
2,302,000,000
2,301,000,000

O₂
O₂
O₂
O₂
O₂
O₂
O₂
O₂
O₂
O₂
O₂
O₂
O₂
O₂
O₂
O₂
O₂
O₂

2.300.000.000
2.299.000.000
2.298.000.000
2.297.000.000
2.296.000.000
2.295.000.000
2.294.000.000
2.293.000.000
2.292.000.000
2.291.000.000
2.290.000.000
2.289.000.000
2.288.000.000
2.287.000.000
2.286.000.000
2.285.000.000
2.284.000.000
2.283.000.000
2.282.000.000
2.281.000.000

2,280,000,000
2,279,000,000
2,278,000,000
2,277,000,000
2,276,000,000
2,275,000,000
2,274,000,000
2,273,000,000
2,272,000,000
2,271,000,000
2,270,000,000
2,269,000,000
2,268,000,000
2,267,000,000
2,266,000,000
2,265,000,000
2,264,000,000
2,263,000,000
2,262,000,000
2,261,000,000

And bit by bit, though tiny we may be,
with many thousands of millennia
to do the job, our gradual release
of excess oxygen into the air
accumulates until, eventually,

2,260,000,000
2,259,000,000
2,258,000,000
2,257,000,000
2,256,000,000
2,255,000,000
2,254,000,000
2,253,000,000
2,252,000,000
2,251,000,000
2,250,000,000
2,249,000,000
2,248,000,000
2,247,000,000
2,246,000,000
2,245,000,000
2,244,000,000
2,243,000,000
2,242,000,000
2,241,000,000

O₂
O₂
O₂
O₂
O₂
O₂
O₂
O₂
O₂
O₂
O₂
O₂
O₂
O₂
O₂

2,240,000,000
2,239,000,000
2,238,000,000
2,237,000,000
2,236,000,000
2,235,000,000
2,234,000,000
2,233,000,000
2,232,000,000
2,231,000,000
2,230,000,000
2,229,000,000
2,228,000,000
2,227,000,000
2,226,000,000
2,225,000,000
2,224,000,000
2,223,000,000
2,222,000,000
2,221,000,000

2,220,000,000 2,219,000,000 2,218,000,000 2,217,000,000 2,216,000,000 2,215,000,000 2,214,000,000 2,213,000,000 2,212,000,000 2,211,000,000 2,210,000,000 2,209,000,000 2,208,000,000 2,207,000,000 2,206,000,000 2,205,000,000 2,204,000,000 2,203,000,000 2,202,000,000 2,201,000,000

2,200,000,000
2,199,000,000
2,198,000,000
2,197,000,000
2,196,000,000
2,195,000,000
2,194,000,000
2,193,000,000
2,192,000,000
2,191,000,000
2,190,000,000
2,189,000,000
2,188,000,000
2,187,000,000
2,186,000,000
2,185,000,000
2,184,000,000
2,183,000,000
2,182,000,000
2,181,000,000

it starts to rust the iron in the rocks.

O₂ O₂ O₂ O₂ O₂ O₂ O₂ O₂ O₂ O₂ O₂ O₂ O₂ O₂ O₂ O₂ O₂ O₂ O₂ O₂ O₂ O₂ O₂ O₂ O₂

2,180,000,000
2,179,000,000
2,178,000,000
2,177,000,000
2,176,000,000
2,175,000,000
2,174,000,000
2,173,000,000
2,172,000,000
2,171,000,000
2,170,000,000
2,169,000,000
2,168,000,000
2,167,000,000
2,166,000,000
2,165,000,000
2,164,000,000
2,163,000,000
2,162,000,000
2,161,000,000

O_2 O_2 O_2 O_2 O_2 O_2 O_2 O_2 O_2 O_2 O_2 O_2 O_2 O_2 O_2 O_2 O_2 O_2 O_2 O_2 O_2 O_2 O_2 O_2 O_2 O_2 O_2 O_2 O_2 O_2 O_2 O_2

2,160,000,000
2,159,000,000
2,158,000,000
2,157,000,000
2,156,000,000
2,155,000,000
2,154,000,000
2,153,000,000
2,152,000,000
2,151,000,000
2,150,000,000
2,149,000,000
2,148,000,000
2,147,000,000
2,146,000,000
2,145,000,000
2,144,000,000
2,143,000,000
2,142,000,000
2,141,000,000

5. EUKARYOTES

It takes two billion years or thereabouts for us prokaryotes to get to where we are today, and still, we're nothing more than strands of RNA and DNA,

2,140,000,000
2,139,000,000
2,138,000,000
2,137,000,000
2,136,000,000
2,135,000,000
2,134,000,000
2,133,000,000
2,132,000,000
2,131,000,000
2,130,000,000
2,129,000,000
2,128,000,000
2,127,000,000
2,126,000,000
2,125,000,000
2,124,000,000
2,123,000,000
2,122,000,000
2,121,000,000

2,120,000,000 2,119,000,000 2,118,000,000 2,117,000,000 2,116,000,000 2,115,000,000 2,114,000,000 2,113,000,000 2,112,000,000 2,111,000,000 2,110,000,000 2,109,000,000 2,108,000,000 2,107,000,000 2,106,000,000 2,105,000,000 2,104,000,000 2,103,000,000 2,102,000,000 2,101,000,

just chemicals inside the tiny cells in which
we share genes, photosynthesize, respire,
and replicate our better qualities.

2,100,000,000
2,099,000,000
2,098,000,000
2,097,000,000
2,096,000,000
2,095,000,000
2,094,000,000
2,093,000,000
2,092,000,000
2,091,000,000
2,090,000,000
2,089,000,000
2,088,000,000
2,087,000,000
2,086,000,000
2,085,000,000
2,084,000,000
2,083,000,000
2,082,000,000
2,081,000,000

2,080,000,000
2,079,000,000
2,078,000,000
2,077,000,000
2,076,000,000
2,075,000,000
2,074,000,000
2,073,000,000
2,072,000,000
2,071,000,000
2,070,000,000
2,069,000,000
2,068,000,000
2,067,000,000
2,066,000,000
2,065,000,000
2,064,000,000
2,063,000,000
2,062,000,000
2,061,000,000

CH₄

2,040,000,000
2,039,000,000
2,038,000,000
2,037,000,000
2,036,000,000
2,035,000,000
2,034,000,000
2,033,000,000
2,032,000,000
2,031,000,000
2,030,000,000
2,029,000,000
2,028,000,000
2,027,000,000
2,026,000,000
2,025,000,000
2,024,000,000
2,023,000,000
2,022,000,000
2,021,000,000

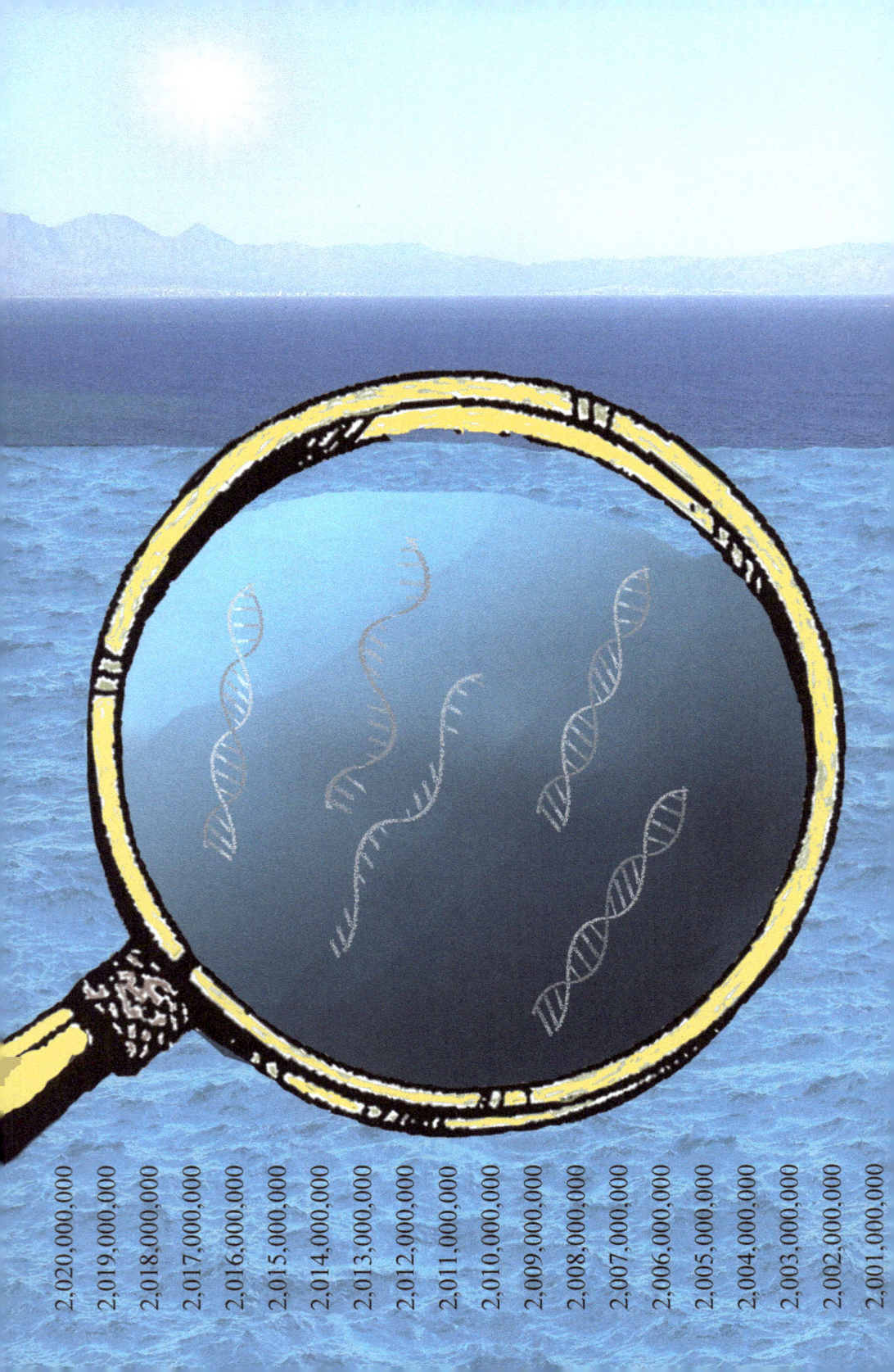

2,020,000,000   2,019,000,000   2,018,000,000   2,017,000,000   2,016,000,000   2,015,000,000   2,014,000,000   2,013,000,000   2,012,000,000   2,011,000,000   2,010,000,000   2,009,000,000   2,008,000,000   2,007,000,000   2,006,000,000   2,005,000,000   2,004,000,000   2,003,000,000   2,002,000,000   2,001,000,000

Some flat, some long, some thick, some spherical,
not one of us is born with arms and legs,
or heads and tails, or hearts, or mouths, or eyes.
We're still just molecules enclosed within
our walls, beset by tides and forces far
beyond our power to control.

2,000,000,000
1,999,000,000
1,998,000,000
1,997,000,000
1,996,000,000
1,995,000,000
1,994,000,000
1,993,000,000
1,992,000,000
1,991,000,000
1,990,000,000
1,989,000,000
1,988,000,000
1,987,000,000
1,986,000,000
1,985,000,000
1,984,000,000
1,983,000,000
1,982,000,000
1,981,000,000

But now,
a day of great significance arrives.
Bacteria that once remained outside,
apart and living independently,
find ways to pierce their neighbors' lipid walls
and make themselves a part of nearby cells.

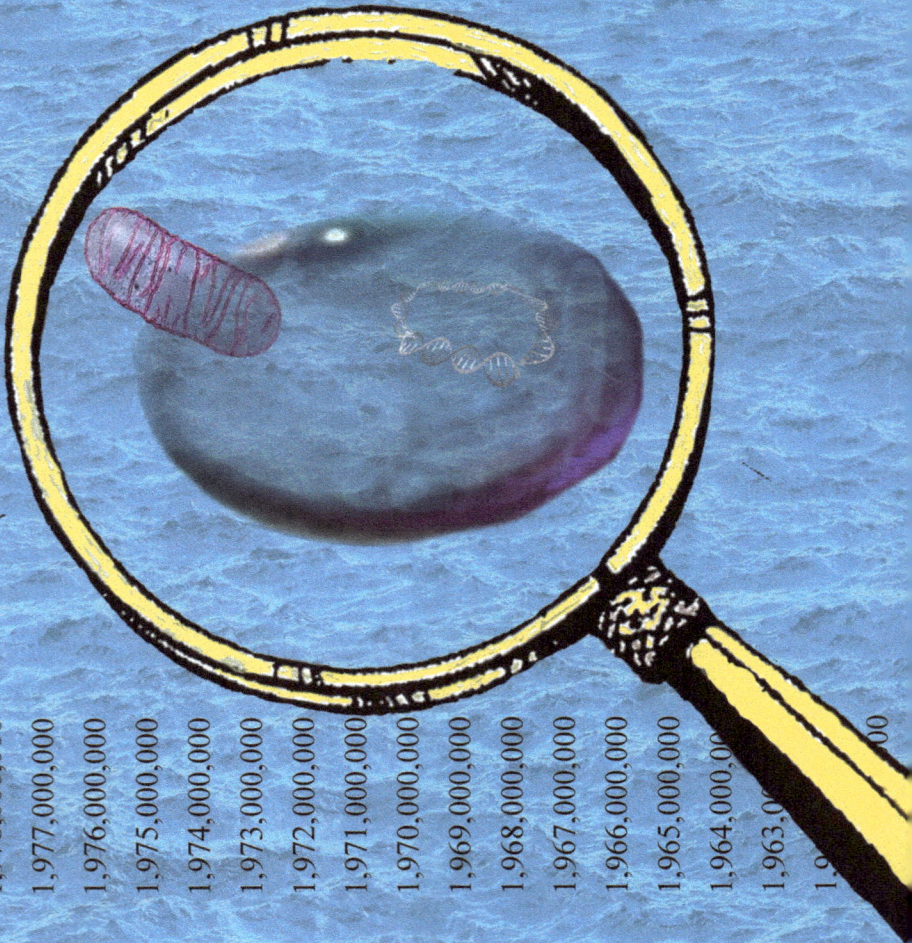

1,980,000,000
1,979,000,000
1,978,000,000
1,977,000,000
1,976,000,000
1,975,000,000
1,974,000,000
1,973,000,000
1,972,000,000
1,971,000,000
1,970,000,000
1,969,000,000
1,968,000,000
1,967,000,000
1,966,000,000
1,965,000,000
1,964,000,0
1,963,0

Thus, some lives join with others. We will call it "endosymbiosis."   With two cells now joined as one... they find new synergy.

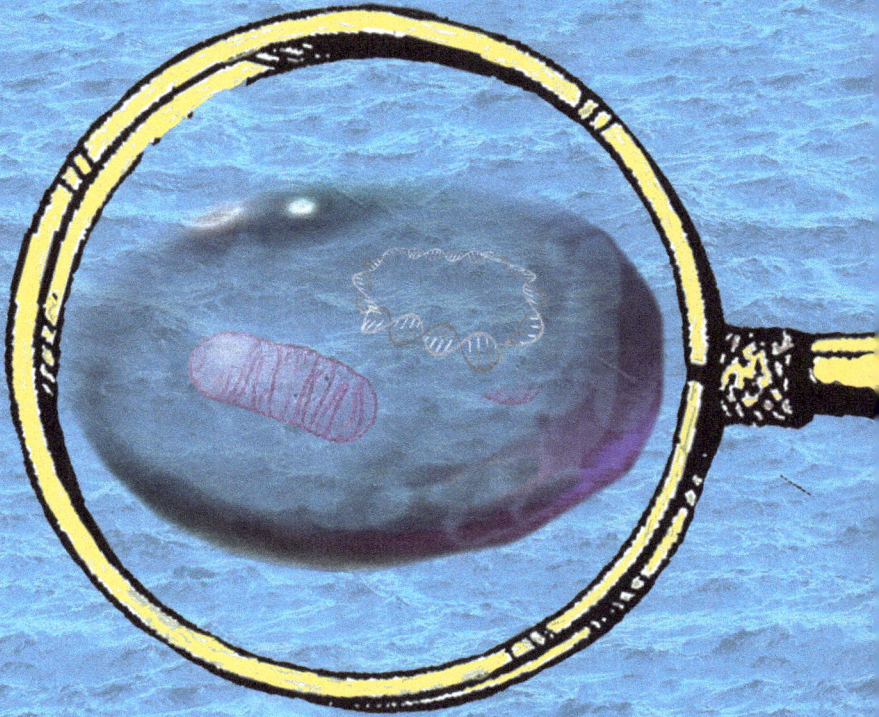

1,960,000,000
1,959,000,000
1,958,000,000
1,957,000,000
1,956,000,000
1,955,000,000
1,954,000,000
1,953,000,000
1,952,000,000
1,951,000,000
1,950,000,000
1,949,000,000
1,948,000,000
1,947,000,000
1,946,000,000
1,945,000,000
1,944,000,000
1,943,000,000
1,942,000,000
1,941,000,000

The way we've always lived begins to change.
Inside their new host cells, bacteria
become what we call mitochondria,
internal power plants, machines that use
the new supply of oxygen to make
adenosine triphosphate — ATP —
and vastly multiply cell energy.

1,940,000,000
1,939,000,000
1,938,000,000
1,937,000,000
1,936,000,000
1,935,000,000
1,934,000,000
1,933,000,000
1,932,000,000
1,931,000,000
1,930,000,000
1,929,000,000
1,928,000,000
1,927,000,000
1,926,000,000
1,925,000,000
1,924,000,000
1,923,000,000
1,922,000,000
1,921,000,000

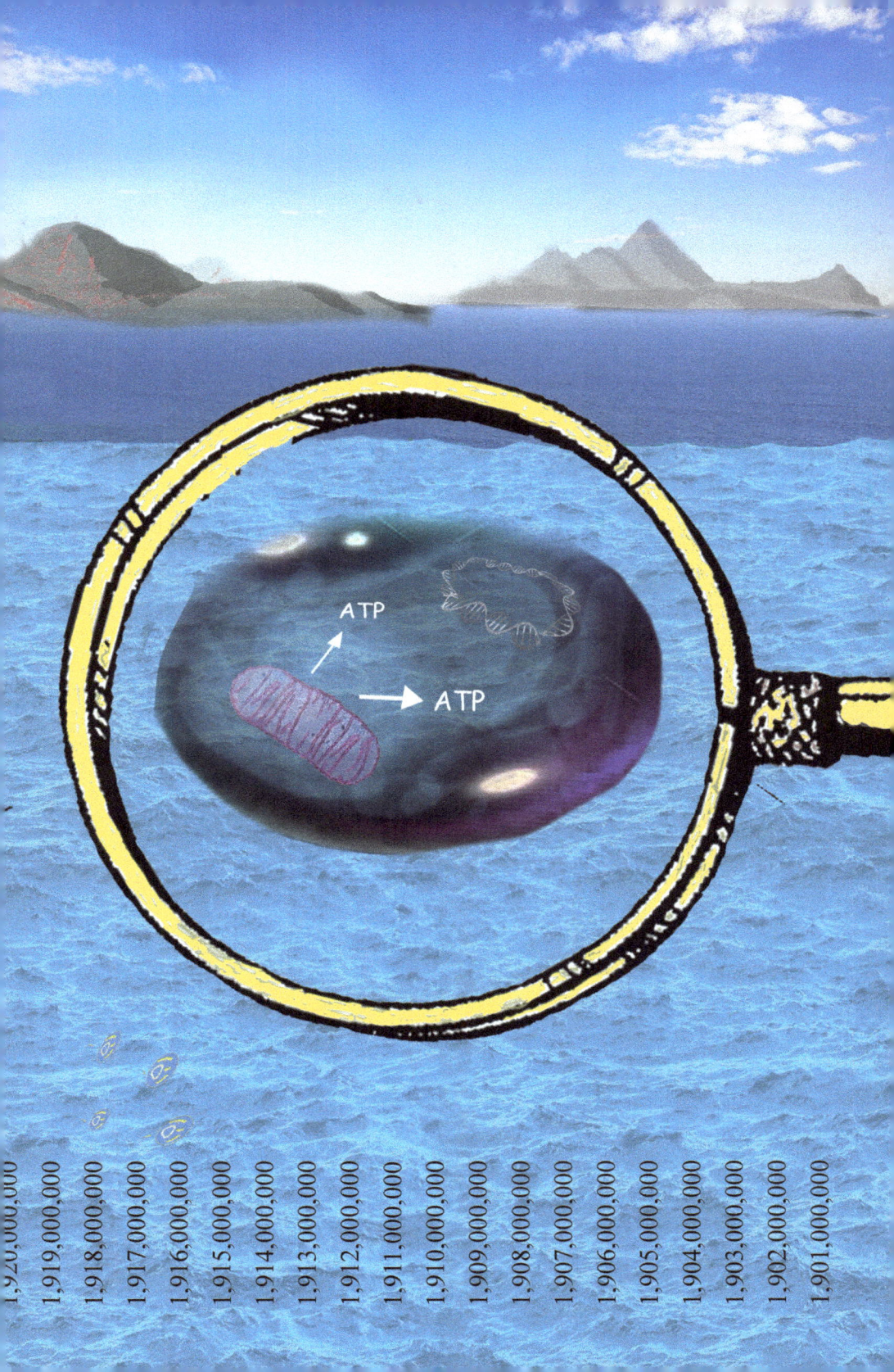

ATP

ATP

1,920,000,000 1,919,000,000 1,918,000,000 1,917,000,000 1,916,000,000 1,915,000,000 1,914,000,000 1,913,000,000 1,912,000,000 1,911,000,000 1,910,000,000 1,909,000,000 1,908,000,000 1,907,000,000 1,906,000,000 1,905,000,000 1,904,000,000 1,903,000,000 1,902,000,000 1,901,000,000

And energized by all that ATP,
cells take into themselves bacteria
adept at making new proteins which, once
inside the cells, we label ribosomes.

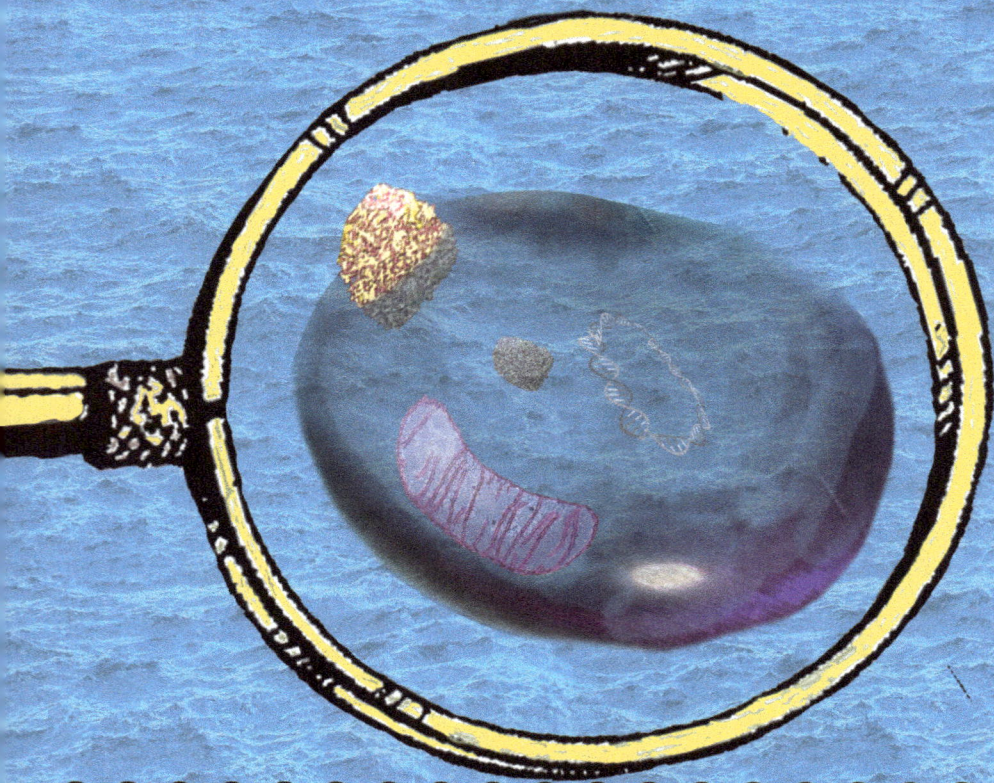

1,900,000,000
1,899,000,000
1,898,000,000
1,897,000,000
1,896,000,000
1,895,000,000
1,894,000,000
1,893,000,000
1,892,000,000
1,891,000,000
1,890,000,000
1,889,000,000
1,888,000,000
1,887,000,000
1,886,000,000
1,885,000,000
1,884,000,000
1,883,000,000
1,882,000,000
1,881,000,000

They're busy things, made up of RNA and proteins. What they do is specialize in reading messages from RNA and using them to make new protein chains.

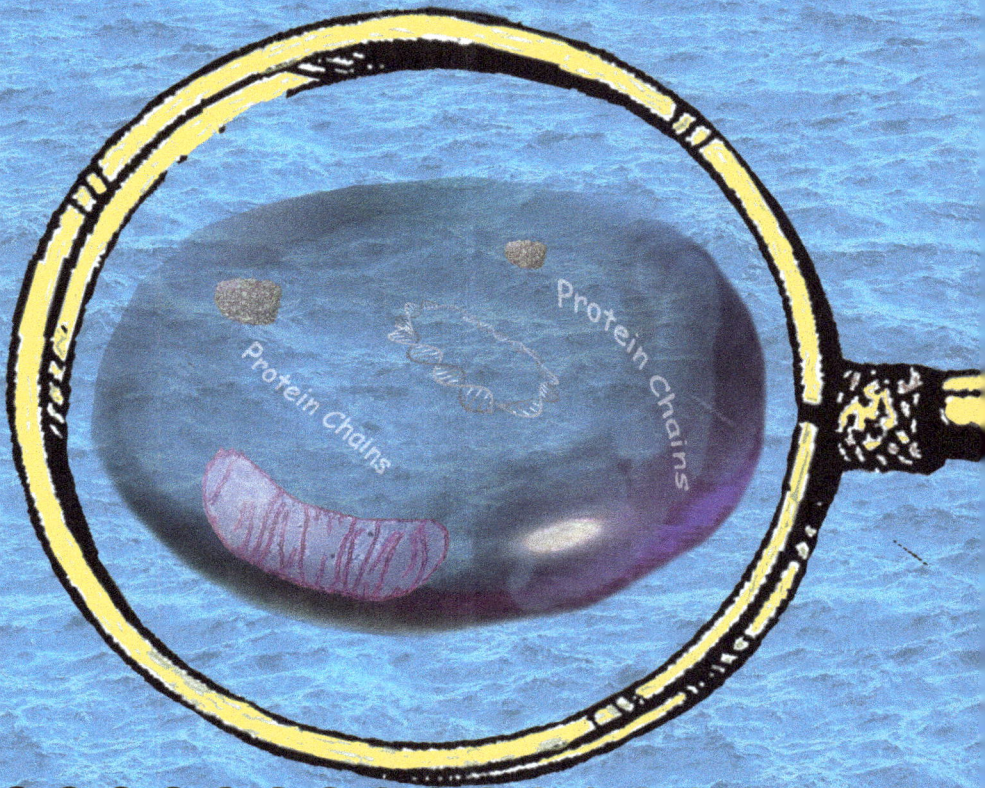

Protein Chains

Protein Chains

1,880,000,000
1,879,000,000
1,878,000,000
1,877,000,000
1,876,000,000
1,875,000,000
1,874,000,000
1,873,000,000
1,872,000,000
1,871,000,000
1,870,000,000
1,869,000,000
1,868,000,000
1,867,000,000
1,866,000,000
1,865,000,000
1,864,000,000
1,863,000,000
1,862,000,000
1,861,000,000

Some of these chains (called "introns") tend to act like scissors, cutting up the other chains, assisting in the process by which all of life evolves.

Intron Protein Chain

1,860,000,000
1,859,000,000
1,858,000,000
1,857,000,000
1,856,000,000
1,855,000,000
1,854,000,000
1,853,000,000
1,852,000,000
1,851,000,000
1,850,000,000
1,849,000,000
1,848,000,000
1,847,000,000
1,846,000,000
1,845,000,000
1,844,000,000
1,843,000,000
1,842,000,000
1,841,000,000

But if the introns cut
up all the DNA that makes us what
we are, we die. And so, as years go by —
a million here, a million there — we learn
to make new lipid membranes to enclose
our DNA, protecting it from all
the frenzy of activity within.

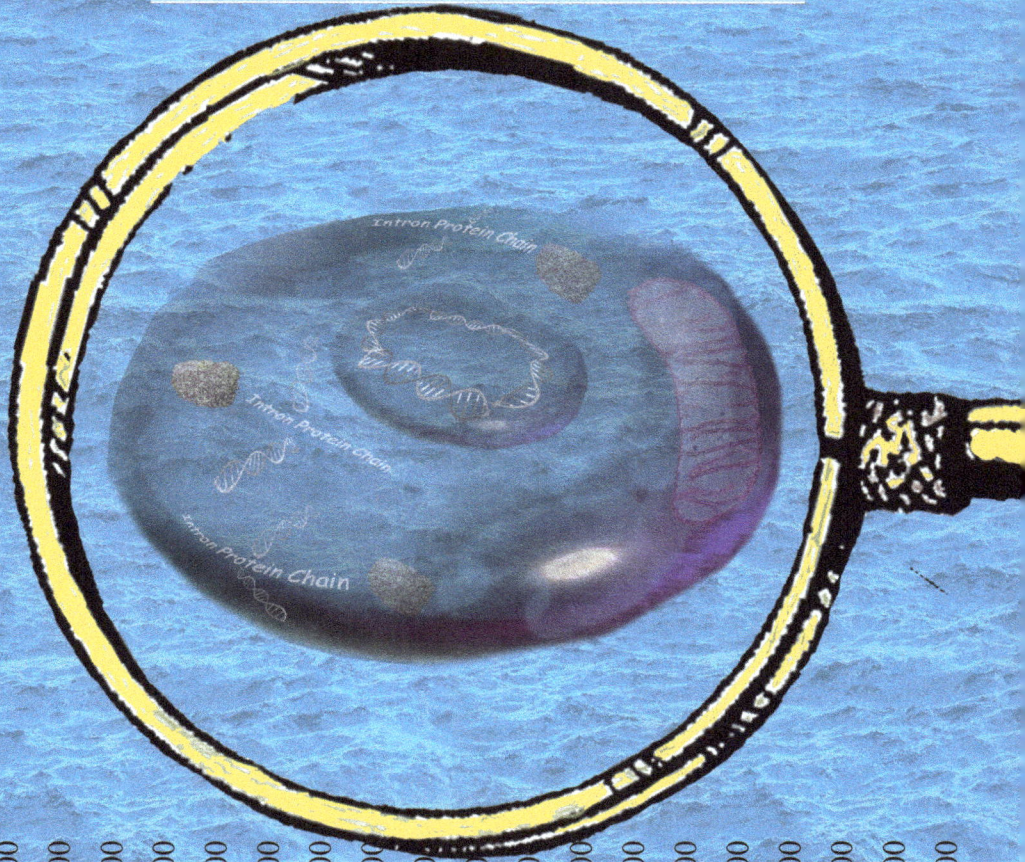

Intron Protein Chain

Intron Protein Chain

Intron Protein Chain

1,840,000,000
1,839,000,000
1,838,000,000
1,837,000,000
1,836,000,000
1,835,000,000
1,834,000,000
1,833,000,000
1,832,000,000
1,831,000,000
1,830,000,000
1,829,000,000
1,828,000,000
1,827,000,000
1,826,000,000
1,825,000,000
1,824,000,000
1,823,000,000
1,822,000,000
1,821,000,000

These new membranes mean we have nuclei,
which means we've now become eukaryotes,
things so complex that life will never be
the same.

Intron Protein Chain

Intron Protein Chain

Intron Protein Chain

First eukaryotes about 1.8 bya

1,820,000,000
1,819,000,000
1,818,000,000
1,817,000,000
1,816,000,000
1,815,000,000
1,814,000,000
1,813,000,000
1,812,000,000
1,811,000,000
1,810,000,000
1,809,000,000
1,808,000,000
1,807,000,000
1,806,000,000
1,805,000,000
1,804,000,000
1,803,000,000
1,802,000,000
1,801,000,00

With all our parts — our nuclei,
our ribosomes, our mitochondria,
our protein chains, our introns, RNA
and DNA, and ATP — each part
a specialist at what it does the best,
we're like a bustling city, full of life,
with energy that's ten or twenty times
the energy of mere prokaryotes.

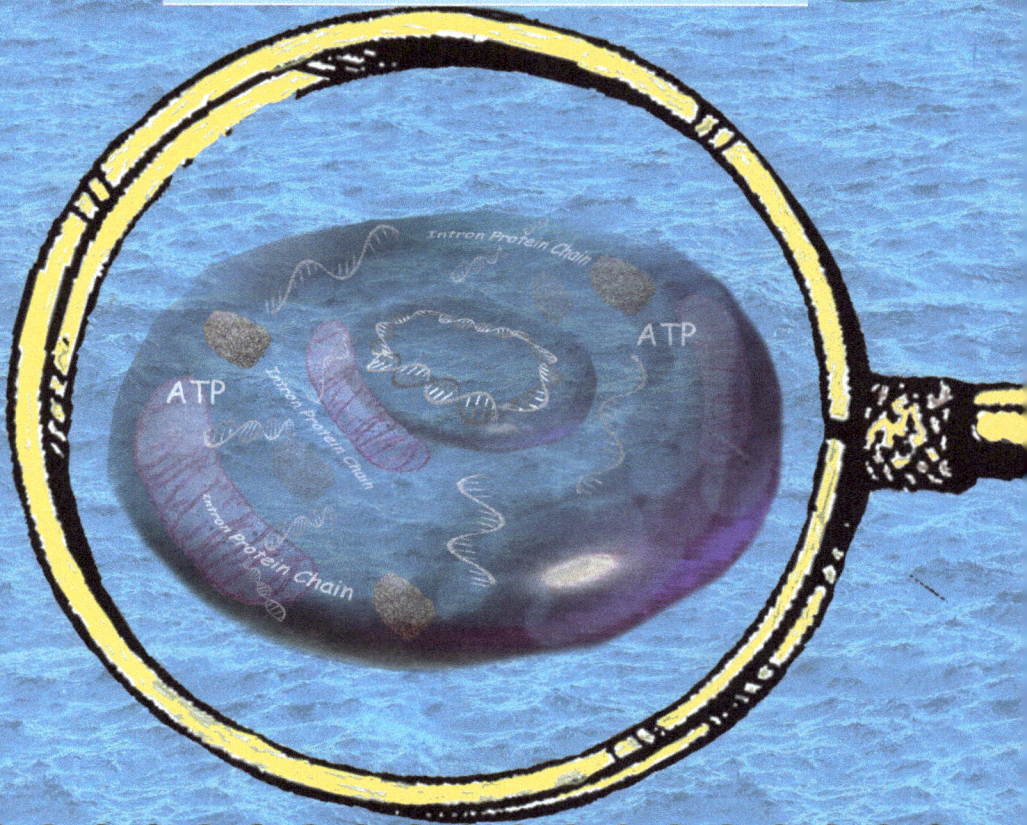

Intron Protein Chain

ATP

ATP

Intron Protein Chain

Intron Protein Chain

1,800,000,000
1,799,000,000
1,798,000,000
1,797,000,000
1,796,000,000
1,795,000,000
1,794,000,000
1,793,000,000
1,792,000,000
1,791,000,000
1,790,000,000
1,789,000,000
1,788,000,000
1,787,000,000
1,786,000,000
1,785,000,000
1,784,000,000
1,783,000,000
1,782,000,000
1,781,000,000

6. DIFFERENTIATION

While older living things continue in the simple ways of life they always have, these cells, with all their organelles, begin to act as if they've found a better way.

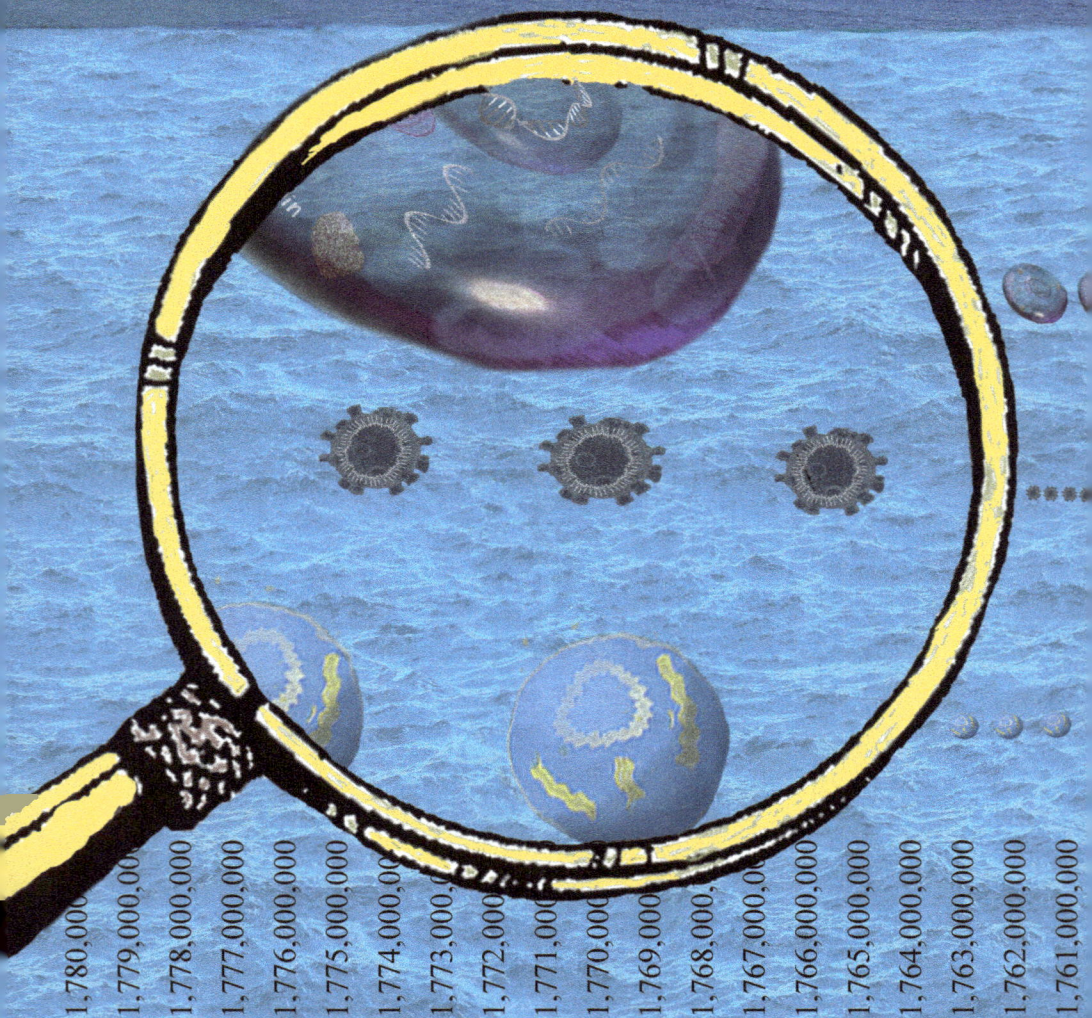

1,780,000,000
1,779,000,000
1,778,000,000
1,777,000,000
1,776,000,000
1,775,000,000
1,774,000,000
1,773,000,000
1,772,000,000
1,771,000,000
1,770,000,000
1,769,000,000
1,768,000,000
1,767,000,000
1,766,000,000
1,765,000,000
1,764,000,000
1,763,000,000
1,762,000,000
1,761,000,000

And so the several forms of life diverge,
with different strategies for how to live.
Will it be viruses, prokaryotes,
or these eukaryotes that prove themselves
to be the ablest forms of life on Earth?

Eukaryotes?

Viruses?

Prokaryotes (Archaea and Bacteria)?

1,760,000,000
1,759,000,000
1,758,000,000
1,757,000,000
1,756,000,000
1,755,000,000
1,754,000,000
1,753,000,000
1,752,000,000
1,751,000,000
1,750,000,000
1,749,000,000
1,748,000,000
1,747,000,000
1,746,000,000
1,745,000,000
1,744,000,000
1,743,000,000
1,742,000,000
1,741,000,000

For viruses, what's simplest seems best, since all they need is RNA to make more copies of themselves, as long as they can find some larger cells to feed upon.

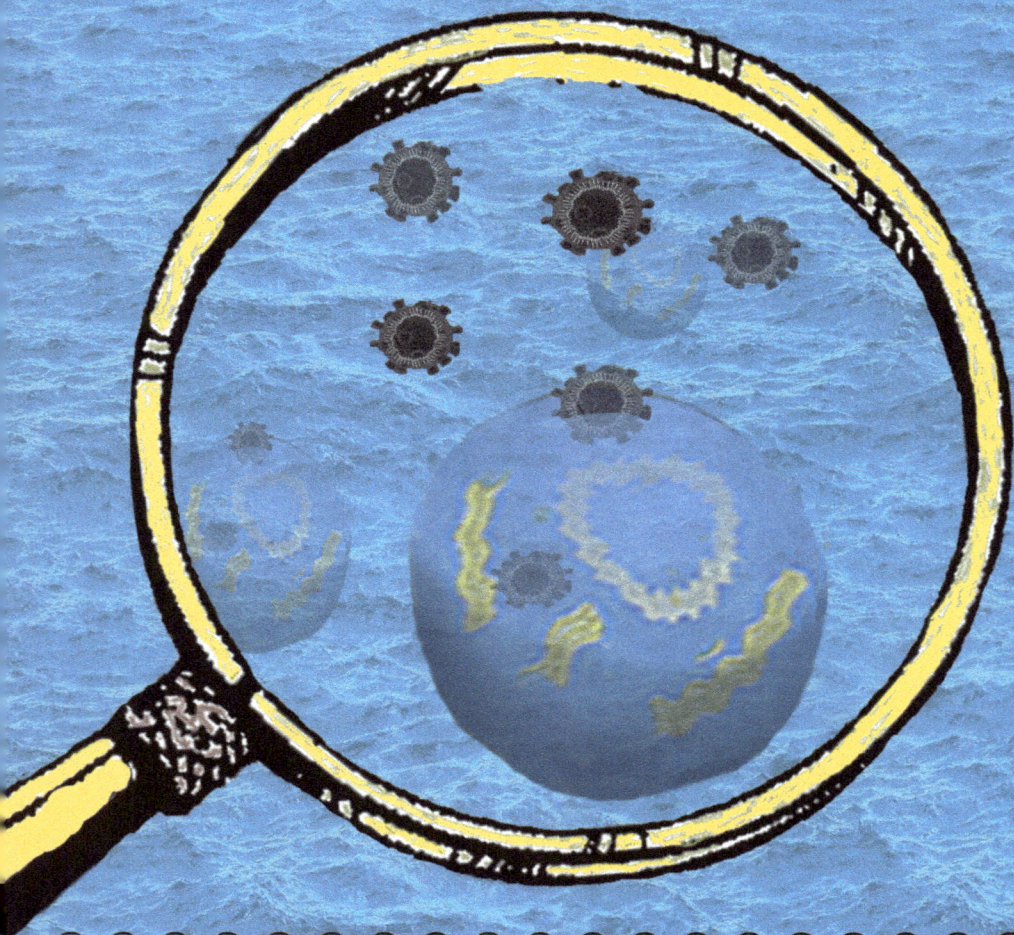

1,740,000,000 1,739,000,000 1,738,000,000 1,737,000,000 1,736,000,000 1,735,000,000 1,734,000,000 1,733,000,000 1,732,000,000 1,731,000,000 1,730,000,000 1,729,000,000 1,728,000,000 1,727,000,000 1,726,000,000 1,725,000,000 1,724,000,000 1,723,000,000 1,722,000,000 1,721,000,000

Prokaryotes lead simple lives as well,
but they're autonomous, each single cell
containing all it needs to carry on,
while counting on mutation and shared genes
to weather any change that comes along.

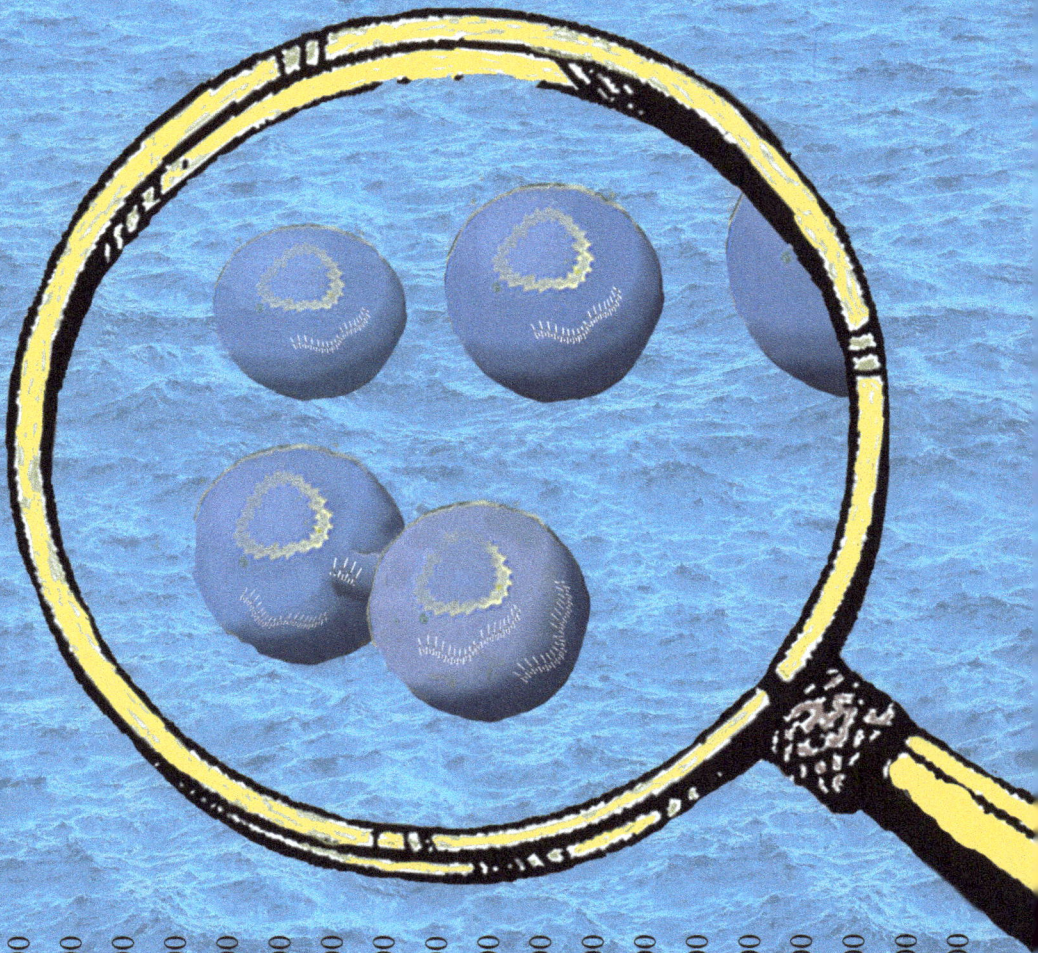

Not so eukaryotes. Committed to
complexity, they try to innovate
in any way they can. A few absorb
sun-loving cyanobacteria,
assimilating them as chloroplasts
and living off their photosynthesis.

Intron Protein Chain

Intron Protein Chain

Intron Protein Chain

1,700,000,000
1,699,000,000
1,698,000,000
1,697,000,000
1,696,000,000
1,695,000,000
1,694,000,000
1,693,000,000
1,692,000,000
1,691,000,000
1,690,000,000
1,689,000,000
1,688,000,000
1,687,000,000
1,686,000,000
1,685,000,000
1,684,000,000
1,683,000,000
1,682,000,000
1,681,000,000

As use of solar power spreads, its waste becomes more prevalent, so that in time more living things start using oxygen to power their internal processes.

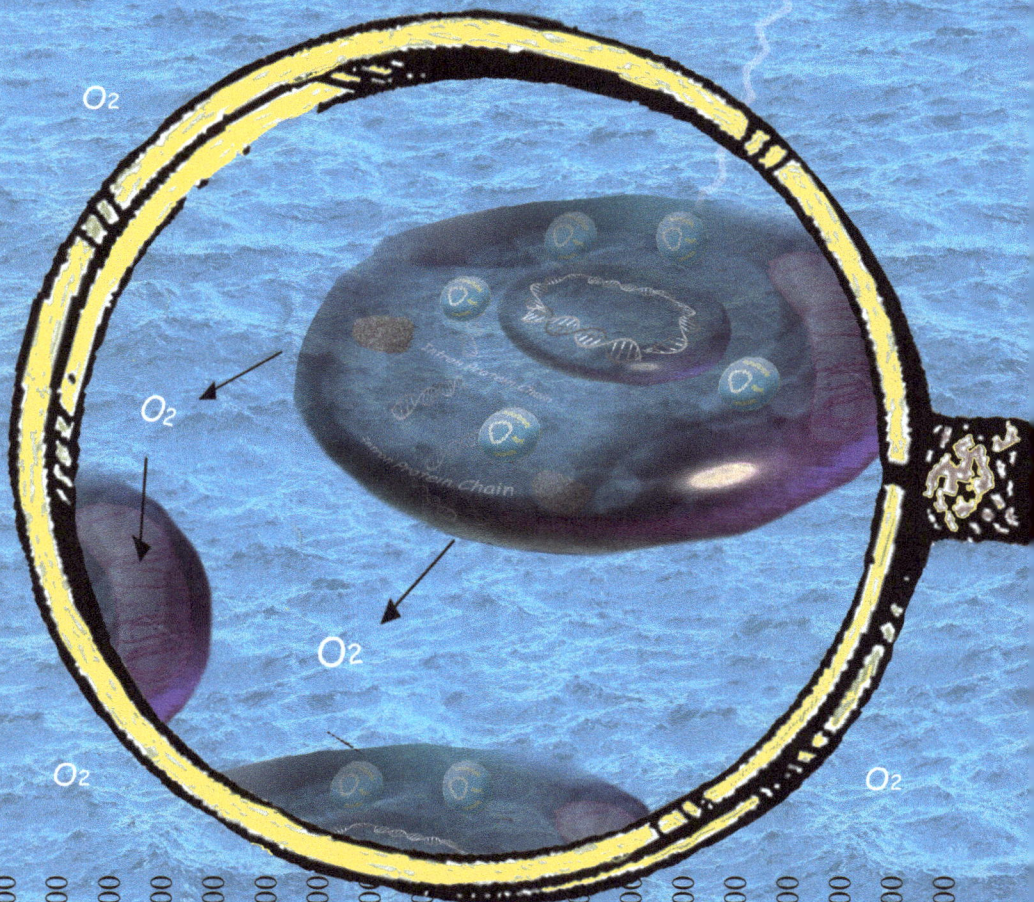

O₂

O₂

O₂

O₂

O₂

O₂

1,680,000,000
1,679,000,000
1,678,000,000
1,677,000,000
1,676,000,000
1,675,000,000
1,674,000,000
1,673,000,000
1,672,000,000
1,671,000,000
1,670,000,000
1,669,000,000
1,668,000,000
1,667,000,000
1,666,000,000
1,665,000,000
1,664,000,000
1,663,000,000
1,662,000,000
1,661,000,000

What's more, with ATP's great energy
and powered by their new fuel, oxygen,
they grow far larger than prokaryotes,
until they're too much for their lipid walls
(which, stretched too thin, collapse under the load).

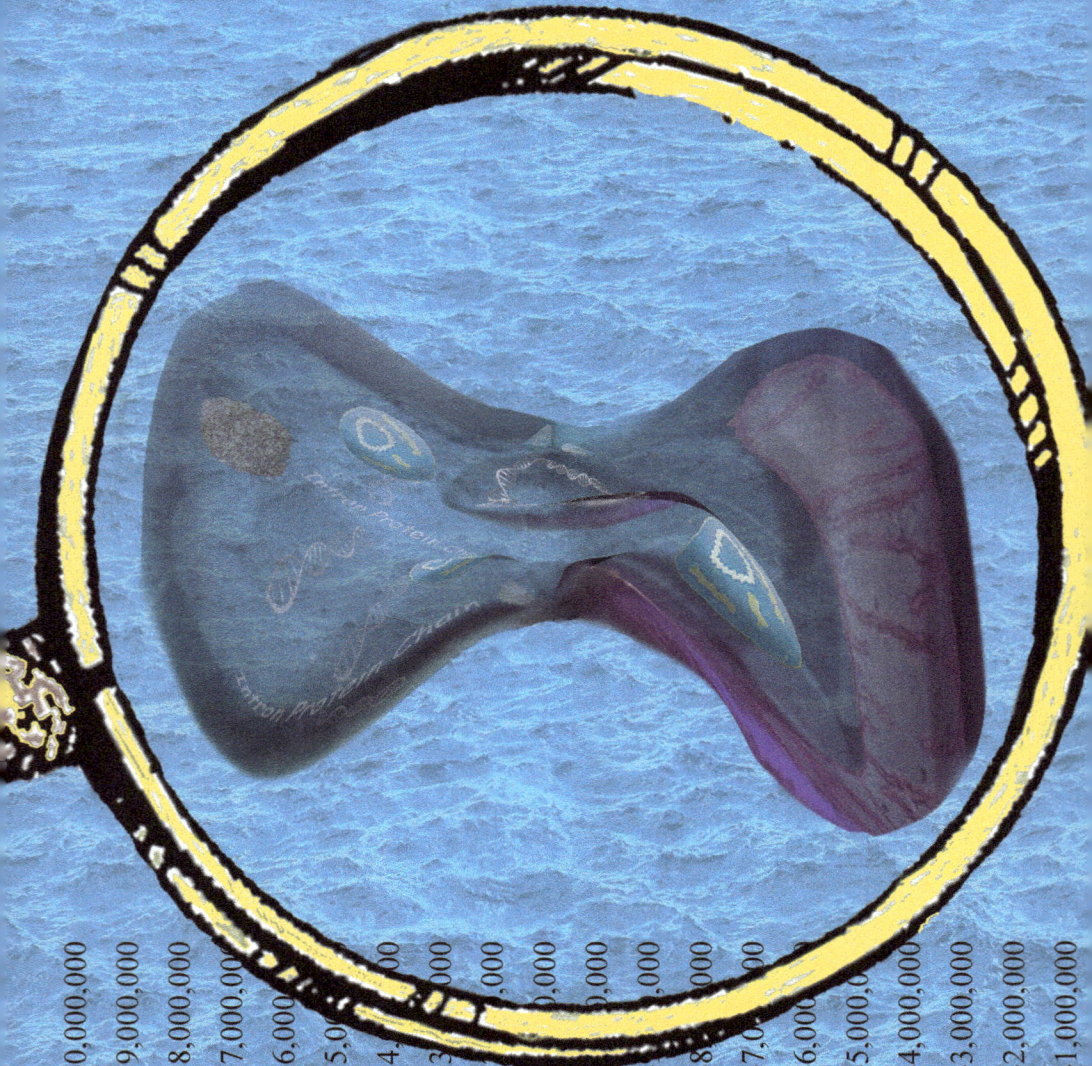

1,660,000,000
1,659,000,000
1,658,000,000
1,657,000,000
1,656,000,000
1,655,000,000
1,654,000,000
1,653,000,000
1,652,000,000
1,651,000,000
1,650,000,000
1,649,000,000
1,648,000,000
1,647,000,000
1,646,000,000
1,645,000,000
1,644,000,000
1,643,000,000
1,642,000,000
1,641,000,000

The cells break up, and even DNA
divides, one perfect copy ending up
in each of the two cells that they become.

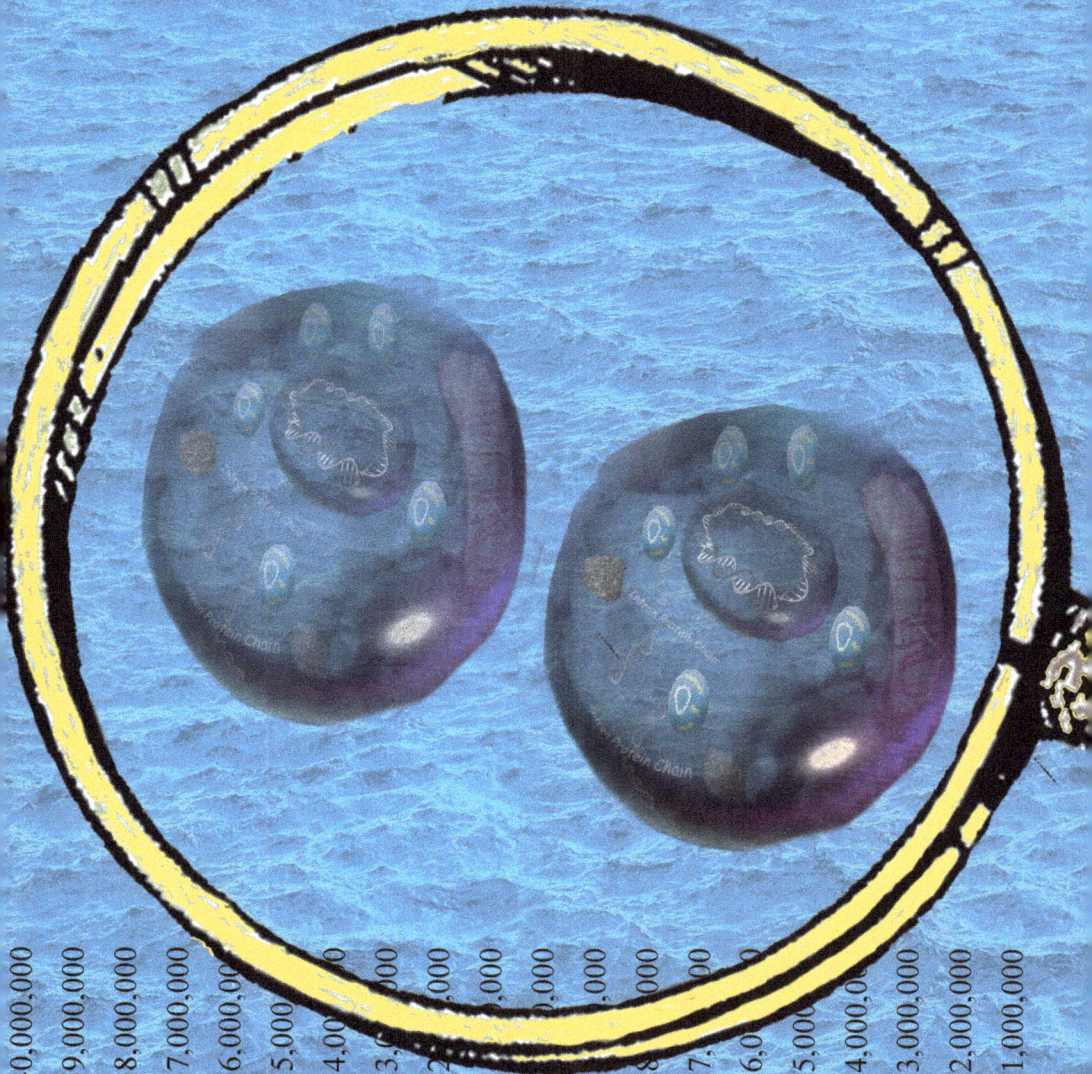

1,640,000,000
1,639,000,000
1,638,000,000
1,637,000,000
1,636,000,000
1,635,000,000
1,634,000,000
1,633,000,000
1,632,000,000
1,63
1,63
1,62
1,628
1,627,
1,626,0
1,625,000,000
1,624,000,0
1,623,000,000
1,622,000,000
1,621,000,000

Now, complex as its chemistry can be,
a single cell can only do so much.

1,620,000,000
1,619,000,000
1,618,000,000
1,617,000,000
1,616,000,000
1,615,000,000
1,614,000,000
1,613,000,000
1,612,000,000
1,611,000,000
1,610,000,000
1,609,000,000
1,608,000,000
1,607,000,000
1,606,000,000
1,605,000,000
1,604,000,000
1,603,000,000
1,602,000,000
1,601,000,000

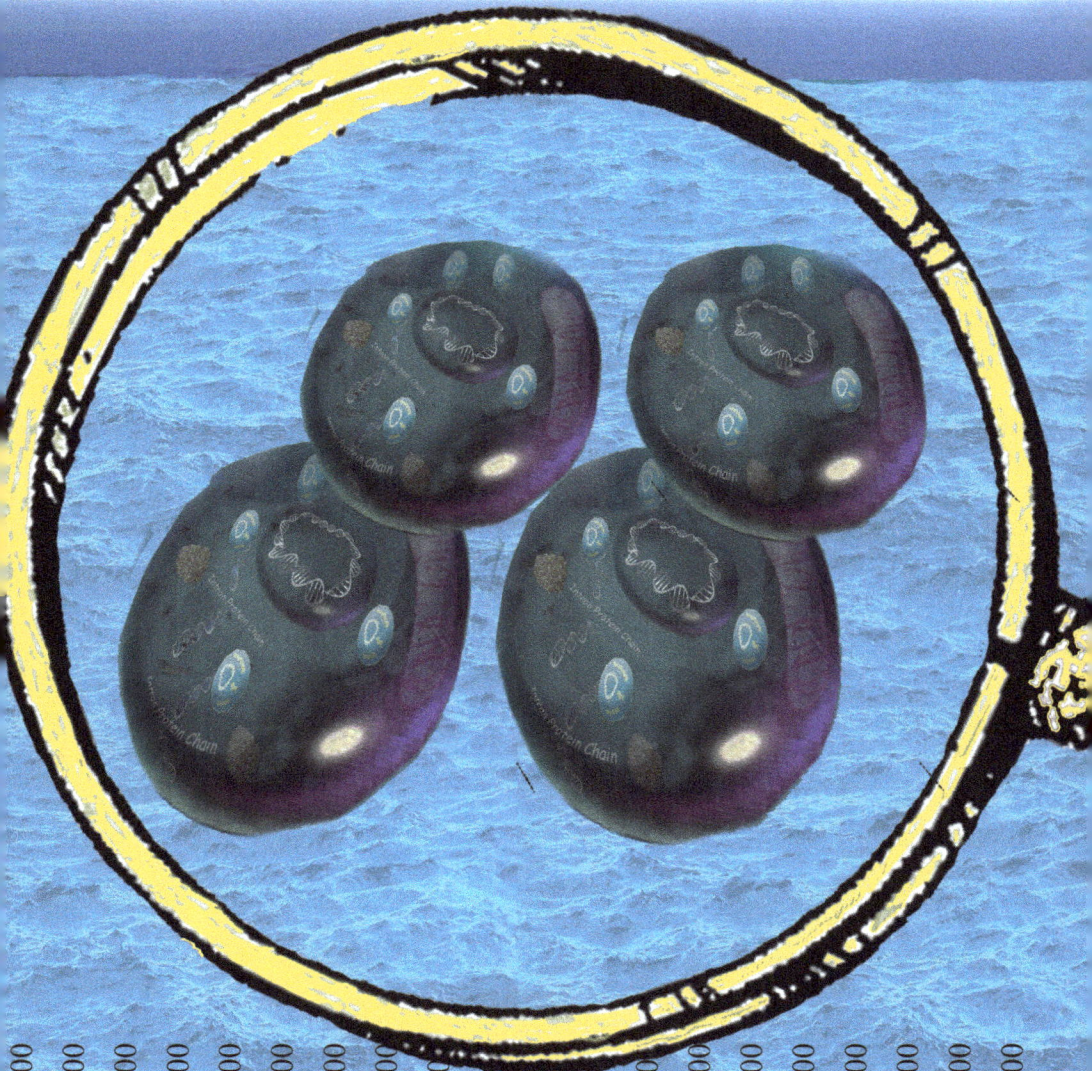

So some eukaryotes try something new:
With DNA all but a perfect match,
and sharing such a lot of other things,
a few dividing cells don't separate.

1,600,000,000
1,599,000,000
1,598,000,000
1,597,000,000
1,596,000,000
1,595,000,000
1,594,000,000
1,593,000,000
1,592,000,000
1,591,000,000
1,590,000,000
1,589,000,000
1,588,000,000
1,587,000,000
1,586,000,000
1,585,000,000
1,584,000,000
1,583,000,000
1,582,000,000
1,581,000,000

They cling to one another, hanging on,
then start to replicate in unison,
becoming complex organisms that
consist of many different kinds of cells,
each kind directed by its DNA
to work together for the common good.

1,580,000,000
1,579,000,000
1,578,000,000
1,577,000,000
1,576,000,000
1,575,000,000
1,574,000,000
1,573,000,000
1,572,000,000
1,571,000,000
1,570,000,000
1,569,000,000
1,568,000,000
1,567,000,000
1,566,000,000
1,565,000,000
1,564,000,000
1,563,000,000
1,562,000,000
1,561,000,000

Some cells form structures good at holding on,
allowing them to anchor well to rocks,

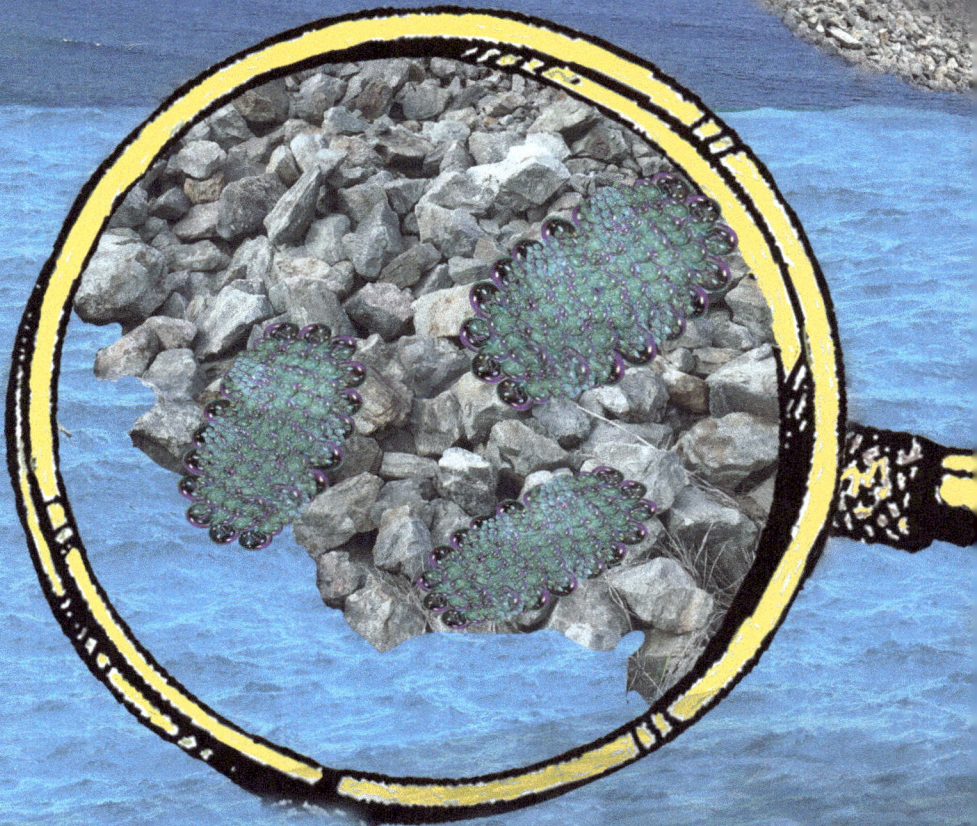

1,560,000,000 1,559,000,000 1,558,000,000 1,557,000,000 1,556,000,000 1,555,000,000 1,554,000,000 1,553,000,000 1,552,000,000 1,551,000,000 1,550,000,000 1,549,000,000 1,548,000,000 1,547,000,000 1,546,000,000 1,545,000,000 1,544,000,000 1,543,000,000 1,542,000,000 1,541,000,000

while others collect pigments that absorb
blue light, which helps them live at greater depths.

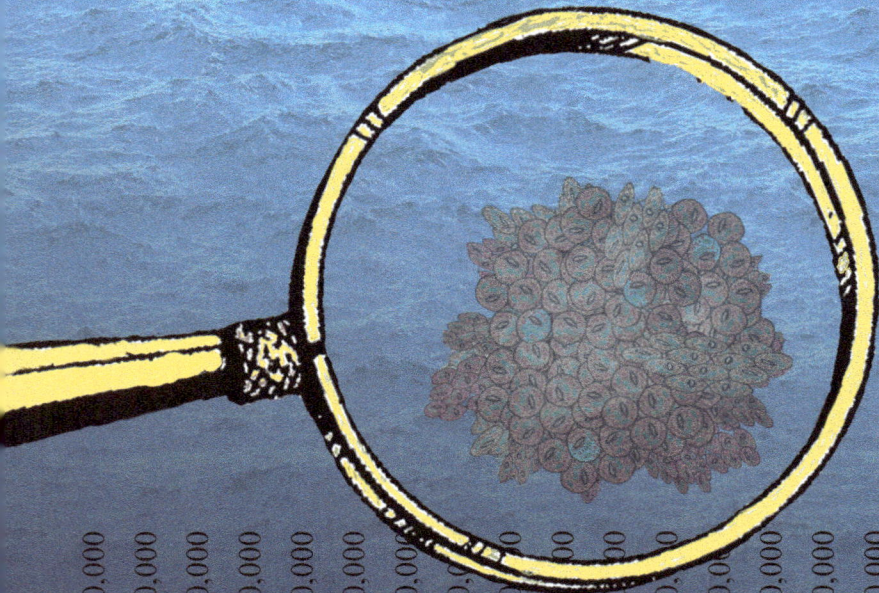

1,540,000,000 1,539,000,000 1,538,000,000 1,537,000,000 1,536,000,000 1,535,000,000 1,534,000,000 1,533,000,000 1,532,000,000 1,531,000,000 1,530,000,000 1,529,000,000 1,528,000,000 1,527,000,000 1,526,000,000 1,525,000,000 1,524,000,000 1,523,000,000 1,522,000,000

Still others specialize in parenting, becoming egg cells that can copy genes from chains of others' DNA, combine them with their own, and make genetically new cells, called zygotes, from the hybrid mix.

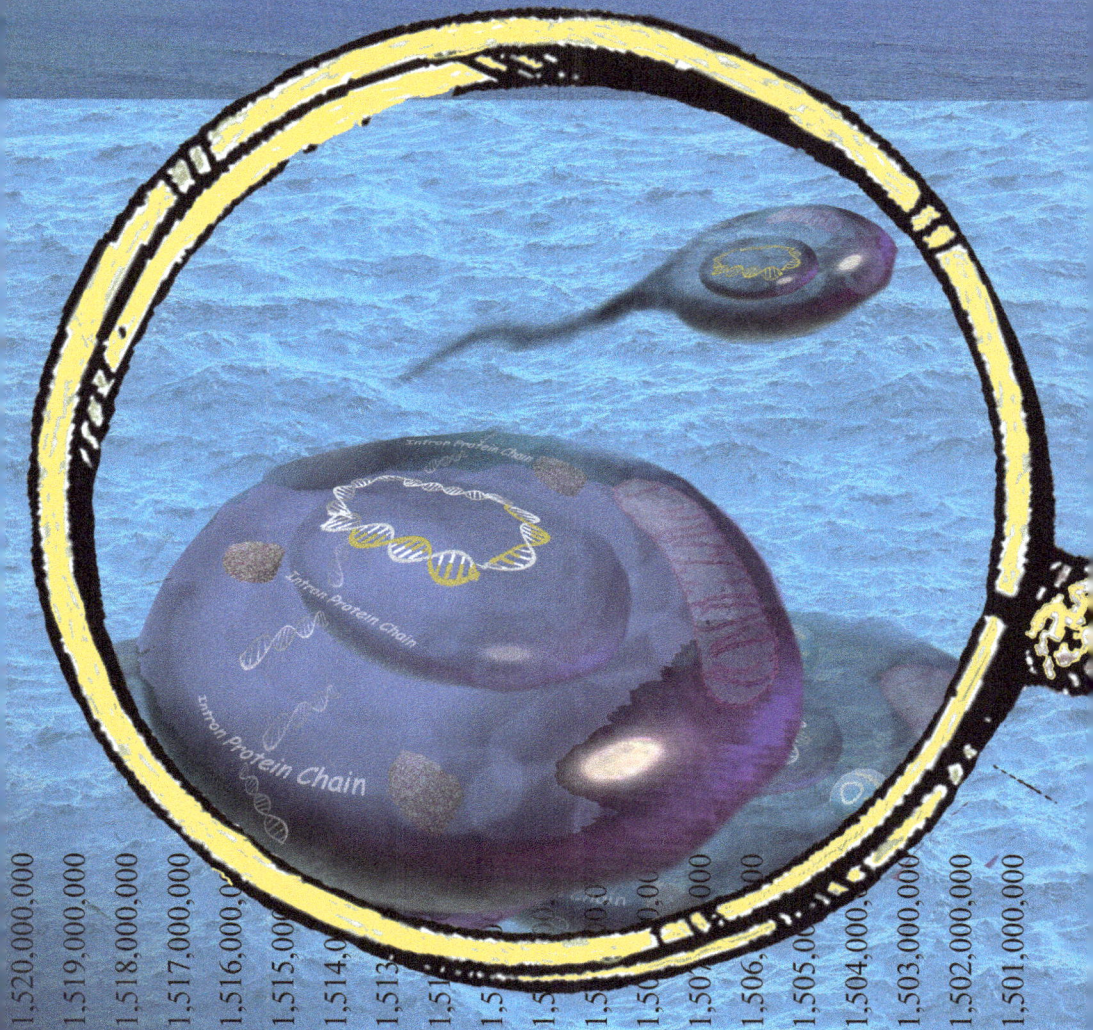

7. ENVIRONMENTAL CHANGE

Now, let's take stock: While we who are alive
have spent two billion years or so on Earth
attempting to adapt, this world of ours
has hardly stayed the same. Sunlight itself
has grown much stronger over time. Some change
has been molecular, the dominance
of sulfur, methane gas and CO_2
diminished by the rise of oxygen.

1,500,000,000
1,499,000,000
1,498,000,000
1,497,000,000
1,496,000,000
1,495,000,000
1,494,000,000
1,493,000,000
1,492,000,000
1,491,000,000
1,490,000,000
1,489,000,000
1,488,000,000
1,487,000,000
1,486,000,000
1,485,000,000
1,484,000,000
1,483,000,000
1,482,000,000
1,481,000,000

1,480,000,000   1,479,000,000   1,478,000,000   1,477,000,000   1,476,000,000   1,475,000,000   1,474,000,000   1,473,000,000   1,472,000,000   1,471,000,000   1,470,000,000   1,469,000,000   1,468,000,000   1,467,000,000   1,466,000,000   1,465,000,000   1,464,000,000   1,463,000,000   1,462,000,000   1,461,000,000

And since the Earth itself is never still,
its crust has been in motion all this time,
its movements pushing mountain ranges up
while sliding coastlines down into the sea.

1,460,000,000 1,459,000,000 1,458,000,000 1,457,000,000 1,456,000,000 1,455,000,000 1,454,000,000 1,453,000,000 1,452,000,000 1,451,000,000 1,450,000,000 1,449,000,000 1,448,000,000 1,447,000,000 1,446,000,000 1,445,000,000 1,444,000,000 1,443,000,000 1,442,000,000 1,441,000,000

So that, across the great expanse of time,
old land masses like Vaalbara have gone,
new supercontinents arisen where
the old ones used to be. Columbia,
the supercontinent that formed almost
a billion years ago, when Vaalbara
had seen its last, now starts to break apart.

1,420,000,000
1,419,000,000
1,418,000,000
1,417,000,000
1,416,000,000
1,415,000,000
1,414,000,000
1,413,000,000
1,412,000,000
1,411,000,000
1,410,000,000
1,409,000,000
1,408,000,000
1,407,000,000
1,406,000,000
1,405,000,000
1,404,000,000
1,403,000,000
1,402,000,000
1,401,000,000

1,400,000,000
1,399,000,000
1,398,000,000
1,397,000,000
1,396,000,000
1,395,000,000
1,394,000,000
1,393,000,000
1,392,000,000
1,391,000,000
1,390,000,000
1,389,000,000
1,388,000,000
1,387,000,000
1,386,000,000
1,385,000,000
1,384,000,000
1,383,000,000
1,382,000,000
1,381,000,000

1,380,000,000
1,379,000,000
1,378,000,000
1,377,000,000
1,376,000,000
1,375,000,000
1,374,000,000
1,373,000,000
1,372,000,000
1,371,000,000
1,370,000,000
1,369,000,000
1,368,000,000
1,367,000,000
1,366,000,000
1,365,000,000
1,364,000,000
1,363,000,000
1,362,000,000
1,361,000,000

The movement of the crust moves water too.
What once were well-lit shallows, inches deep,
have now been so well flooded that good light
is hard to find.

1,360,000,000
1,359,000,000
1,358,000,000
1,357,000,000
1,356,000,000
1,355,000,000
1,354,000,000
1,353,000,000
1,352,000,000
1,351,000,000
1,350,000,000
1,349,000,000
1,348,000,000
1,347,000,000
1,346,000,000
1,345,000,000
1,344,000,000
1,343,000,000
1,342,000,000
1,341,000,000

1,340,000,000
1,339,000,000
1,338,000,000
1,337,000,000
1,336,000,000
1,335,000,000
1,334,000,000
1,333,000,000
1,332,000,000
1,331,000,000
1,330,000,000
1,329,000,000
1,328,000,000
1,327,000,000
1,326,000,000
1,325,000,000
1,324,000,000
1,323,000,000
1,322,000,000
1,321,000,000

1,320,000,000
1,319,000,000
1,318,000,000
1,317,000,000
1,316,000,000
1,315,000,000
1,314,000,000
1,313,000,000
1,312,000,000
1,311,000,000
1,310,000,000
1,309,000,000
1,308,000,000
1,307,000,000
1,306,000,000
1,305,000,000
1,304,000,000
1,303,000,000
1,302,000,000
1,301,000,000

But keep in mind that when
the Earth's crust shifts, the continents advance
just centimeters every year. And when
such massive continents take that much time
to move an inch, their progress can seem slow
to those whose lives last less than centuries.

1,300,000,000
1,299,000,000
1,298,000,000
1,297,000,000
1,296,000,000
1,295,000,000
1,294,000,000
1,293,000,000
1,292,000,000
1,291,000,000
1,290,000,000
1,289,000,000
1,288,000,000
1,287,000,000
1,286,000,000
1,285,000,000
1,284,000,000
1,283,000,000
1,282,000,000
1,281,000,000

1,280,000,000
1,279,000,000
1,278,000,000
1,277,000,000
1,276,000,000
1,275,000,000
1,274,000,000
1,273,000,000
1,272,000,000
1,271,000,000
1,270,000,000
1,269,000,000
1,268,000,000
1,267,000,000
1,266,000,000
1,265,000,000
1,264,000,000
1,263,000,000
1,262,000,000
1,261,000,000

All told, we living things have ample time to acclimate ourselves to all the ways that changes in the world affect our lives.

1,260,000,000
1,259,000,000
1,258,000,000
1,257,000,000
1,256,000,000
1,255,000,000
1,254,000,000
1,253,000,000
1,252,000,000
1,251,000,000
1,250,000,000
1,249,000,000
1,248,000,000
1,247,000,000
1,246,000,000
1,245,000,000
1,244,000,000
1,243,000,000
1,242,000,000
1,241,000,000

1,240,000,000 1,239,000,000 1,238,000,000 1,237,000,000 1,236,000,000 1,235,000,000 1,234,000,000 1,233,000,000 1,232,000,000 1,231,000,000 1,230,000,000 1,229,000,000 1,228,000,000 1,227,000,000 1,226,000,000 1,225,000,000 1,224,000,000 1,223,000,000 1,222,000,000 1,221,000,000

Sometimes it seems as if the continents
are just too vast, and change too gradual,
for lives as insignificant as ours.

1,220,000,000
1,219,000,000
1,218,000,000
1,217,000,000
1,216,000,000
1,215,000,000
1,214,000,000
1,213,000,000
1,212,000,000
1,211,000,000
1,210,000,000
1,209,000,000
1,208,000,000
1,207,000,000
1,206,000,000
1,205,000,000
1,204,000,000
1,203,000,000
1,202,000,000
1,201,000,000

1,200,000,000
1,199,000,000
1,198,000,000
1,197,000,000
1,196,000,000
1,195,000,000
1,194,000,000
1,193,000,000
1,192,000,000
1,191,000,000
1,190,000,000
1,189,000,000
1,188,000,000
1,187,000,000
1,186,000,000
1,185,000,000
1,184,000,000
1,183,000,000
1,182,000,000
1,181,000,000

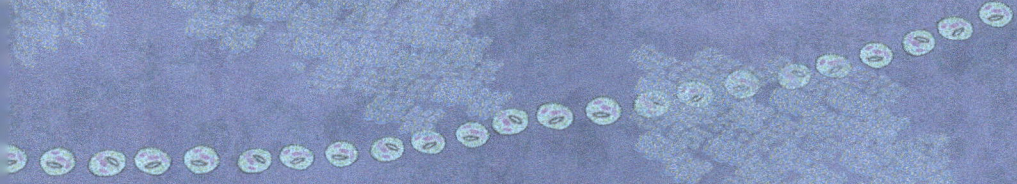

But changes do take place: whole continents
divide and separate; the chemicals
that fill the oceans change; the atmosphere
does too; rocks oxidize; the sun grows strong.

1,180,000,000
1,179,000,000
1,178,000,000
1,177,000,000
1,176,000,000
1,175,000,000
1,174,000,000
1,173,000,000
1,172,000,000
1,171,000,000
1,170,000,000
1,169,000,000
1,168,000,000
1,167,000,000
1,166,000,000
1,165,000,000
1,164,000,000
1,163,000,000
1,162,000,000
1,161,000,000

1,160,000,000
1,159,000,000
1,158,000,000
1,157,000,000
1,156,000,000
1,155,000,000
1,154,000,000
1,153,000,000
1,152,000,000
1,151,000,000
1,150,000,000
1,149,000,000
1,148,000,000
1,147,000,000
1,146,000,000
1,145,000,000
1,144,000,000
1,143,000,000
1,142,000,000
1,141,000,000

O₂ O₂ O₂ O₂ O₂

Supercontinent Atlantica breaks up,
Rodinia begins to form, about 1.13 bya

1,140,000,000
1,139,000,000
1,138,000,000
1,137,000,000
1,136,000,000
1,135,000,000
1,134,000,000
1,133,000,000
1,132,000,000
1,131,000,000
1,130,000,000
1,129,000,000
1,128,000,000
1,127,000,000
1,126,000,000
1,125,000,000
1,124,000,000
1,123,000,000
1,122,000,000
1,121,000,000

1,120,000,000

1,119,000,000

1,118,000,000

1,117,000,000

1,116,000,000

1,115,000,000

1,114,000,000

1,113,000,000

1,112,000,000

1,111,000,000

1,110,000,000

1,109,000,000

1,108,000,000

1,107,000,000

1,106,000,000

1,105,000,000

1,104,000,000

1,103,000,000

1,102,000,000

1,101,000,000

8. TIME

O2
O2
O2
O2
O2

The good news is that all this change takes time.
We had a hundred million years or so
to fashion RNA from acid chains,
a billion years, or thereabouts, to find
the keys to photosynthesis.

PO4
C5H10O5 — C5H5N5
PO4
C5H10O5 — C4H5N3O2
PO4
C5H10O5 — C4H5N3O
PO4
C5H10O5 — C5H5N5O
PO4
C5H10O5 — C4H5N3O
PO4
C5H10O5 — C4H5N3O

PO4
C5H10O5 — C5H5N5O

PO4
|
C5H10O5 — C4H4N2O2

PO4
|
C5H10O5

PO4
C5H10O5 — C4H5N5

1,100,000,000
1,099,000,000
1,098,000,000
1,097,000,000
1,096,000,000
1,095,000,000
1,094,000,000
1,093,000,000
1,092,000,000
1,091,000,000
1,090,000,000
1,089,000,000
1,088,000,000
1,087,000,000
1,086,000,000
1,085,000,000
1,084,000,000
1,083,000,000
1,082,000,000
1,081,000,000

O₂ O₂ O₂ O₂ O₂ O₂ O₂ O₂ O₂ O₂ O₂ O₂ O₂ O₂ O₂ O₂

1,080,000,000
1,079,000,000
1,078,000,000
1,077,000,000
1,076,000,000
1,075,000,000
1,074,000,000
1,073,000,000
1,072,000,000
1,071,000,000
1,070,000,000
1,069,000,000
1,068,000,000
1,067,000,000
1,066,000,000
1,065,000,000
1,064,000,000
1,063,000,000
1,062,000,000
1,061,000,000

And now —
about two billion more — we're finally
adjusting to a world of oxygen.

1,060,000,000 1,059,000,000 1,058,000,000 1,057,000,000 1,056,000,000 1,055,000,000 1,054,000,000 1,053,000,000 1,052,000,000 1,051,000,000 1,050,000,000 1,049,000,000 1,048,000,000 1,047,000,000 1,046,000,000 1,045,000,000 1,044,000,000 1,043,000,000 1,042,000,000 1,041,000,000

1,040,000,000
1,039,000,000
1,038,000,000
1,037,000,000
1,036,000,000
1,035,000,000
1,034,000,000
1,033,000,000
1,032,000,000
1,031,000,000
1,030,000,000
1,029,000,000
1,028,000,000
1,027,000,000
1,026,000,000
1,025,000,000
1,024,000,000
1,023,000,000
1,022,000,000
1,021,000,000

So far, we've always had sufficient time to find the right nucleotides, the right genetic codes, to fit the changing world.

1,020,000,000
1,019,000,000
1,018,000,000
1,017,000,000
1,016,000,000
1,015,000,000
1,014,000,000
1,013,000,000
1,012,000,000
1,011,000,000
1,010,000,000
1,009,000,000
1,008,000,000
1,007,000,000
1,006,000,000
1,005,000,000
1,004,000,000
1,003,000,000
1,002,000,000
1,001,000,000

O2
O2
O2
O2
O2
O2
O2
O2
O2
O2
O2
O2
O2
O2
O2
O2
O2
O2
O2
O2
O2
O2

1,000,000,000
999,000,000
998,000,000
997,000,000
996,000,000
995,000,000
994,000,000
993,000,000
992,000,000
991,000,000
990,000,000
989,000,000
988,000,000
987,000,000
986,000,000
985,000,000
984,000,000
983,000,000
982,000,000
981,000,000

Yet if we were to try to read this book —
not only all the lines we're reading now,
but those across the bottom of each page,
"nine hundred eighty million" and the rest —
could we recite four billion years out loud,
a million at a time?

980,000,000 979,000,000 978,000,000 977,000,000 976,000,000 975,000,000 974,000,000 973,000,000 972,000,000 971,000,000 970,000,000 969,000,000 968,000,000 967,000,000 966,000,000 965,000,000 964,000,000 963,000,000 962,000,000 961,000,000

Unlikely, sure...
Since we're eukaryotes, we're not inclined
to take on tasks requiring so much time.

960,000,000
959,000,000
958,000,000
957,000,000
956,000,000
955,000,000
954,000,000
953,000,000
952,000,000
951,000,000
950,000,000
949,000,000
948,000,000
947,000,000
946,000,000
945,000,000
944,000,000
943,000,000
942,000,000
941,000,000

940,000,000
939,000,000
938,000,000
937,000,000
936,000,000
935,000,000
934,000,000
933,000,000
932,000,000
931,000,000
930,000,000
929,000,000
928,000,000
927,000,000
926,000,000
925,000,000
924,000,000
923,000,000
922,000,000
921,000,000

1 million years

∧

920,000,000 919,000,000 918,000,000 917,000,000 916,000,000 915,000,000 914,000,000 913,000,000 912,000,000 911,000,000 910,000,000 909,000,000 908,000,000 907,000,000 906,000,000 905,000,000 904,000,000 903,000,000 902,000,000 901,000,000

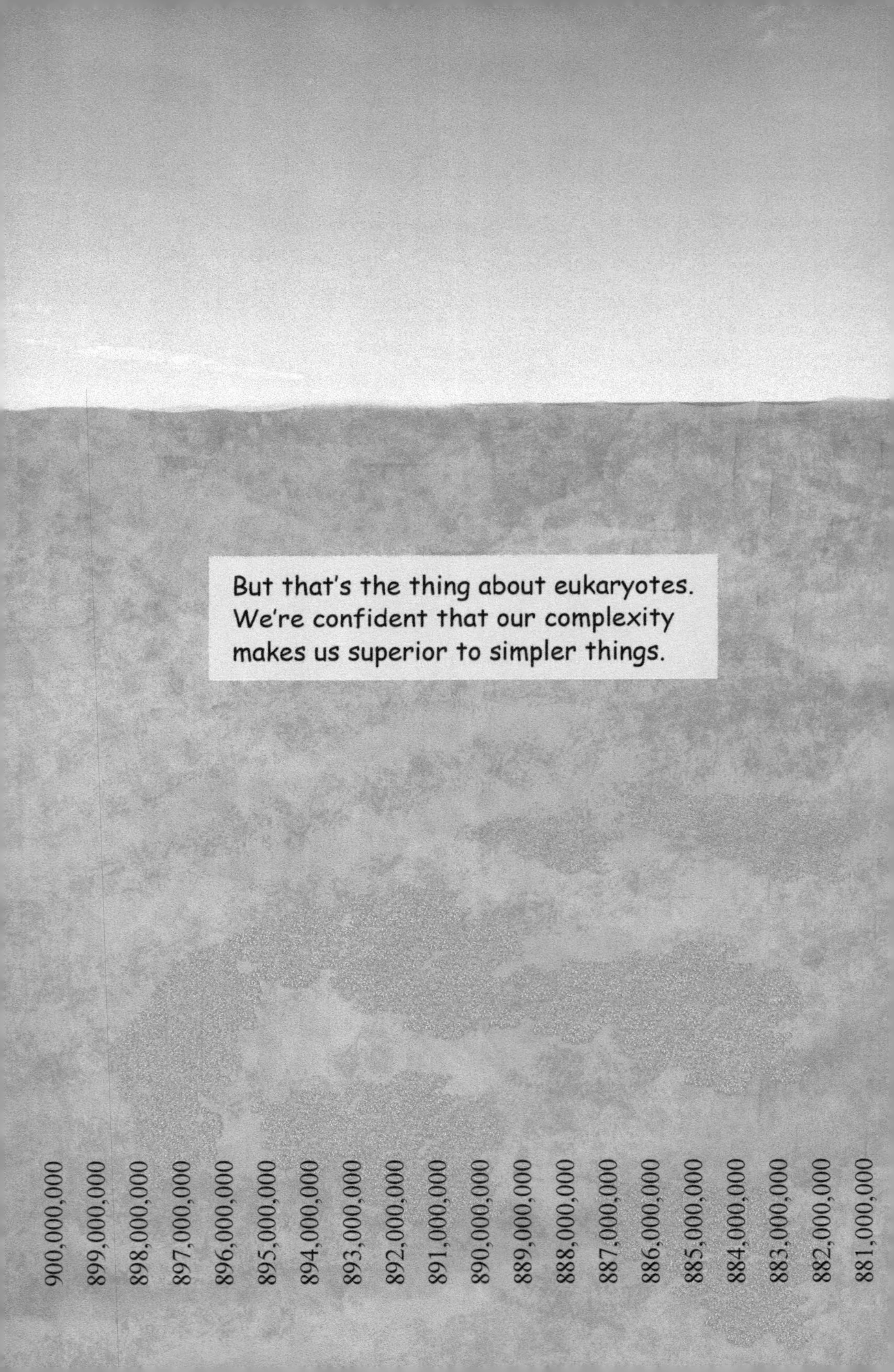

But that's the thing about eukaryotes.
We're confident that our complexity
makes us superior to simpler things.

900,000,000

899,000,000

898,000,000

897,000,000

896,000,000

895,000,000

894,000,000

893,000,000

892,000,000

891,000,000

890,000,000

889,000,000

888,000,000

887,000,000

886,000,000

885,000,000

884,000,000

883,000,000

882,000,000

881,000,000

Intron Protein Chain

Intron Protein Chain

Intron Protein Chain

880,000,000 879,000,000 878,000,000 877,000,000 876,000,000 875,000,000 874,000,000 873,000,000 872,000,000 871,000,000 870,000,000 869,000,000 868,000,000 867,000,000 866,000,000 865,000,000 864,000,000 863,000,000 862,000,000 861,000,000

Convinced we've found a better way of life,
we're not content with things the way they are.
We're always looking for new things to do.

840,000,000
839,000,000
838,000,000
837,000,000
836,000,000
835,000,000
834,000,000
833,000,000
832,000,000
831,000,000
830,000,000
829,000,000
828,000,000
827,000,000
826,000,000
825,000,000
824,000,000
823,000,000
822,000,000
821,000,000

And so we look to dry Rodinia, the newest supercontinent, and we set out to colonize its shores as well.

820,000,000 819,000,000 818,000,000 817,000,000 816,000,000 815,000,000 814,000,000 813,000,000 812,000,000 811,000,000 810,000,000 809,000,000 808,000,000 807,000,000 806,000,000 805,000,000 804,000,000 803,000,000 802,000,000 801,000,000

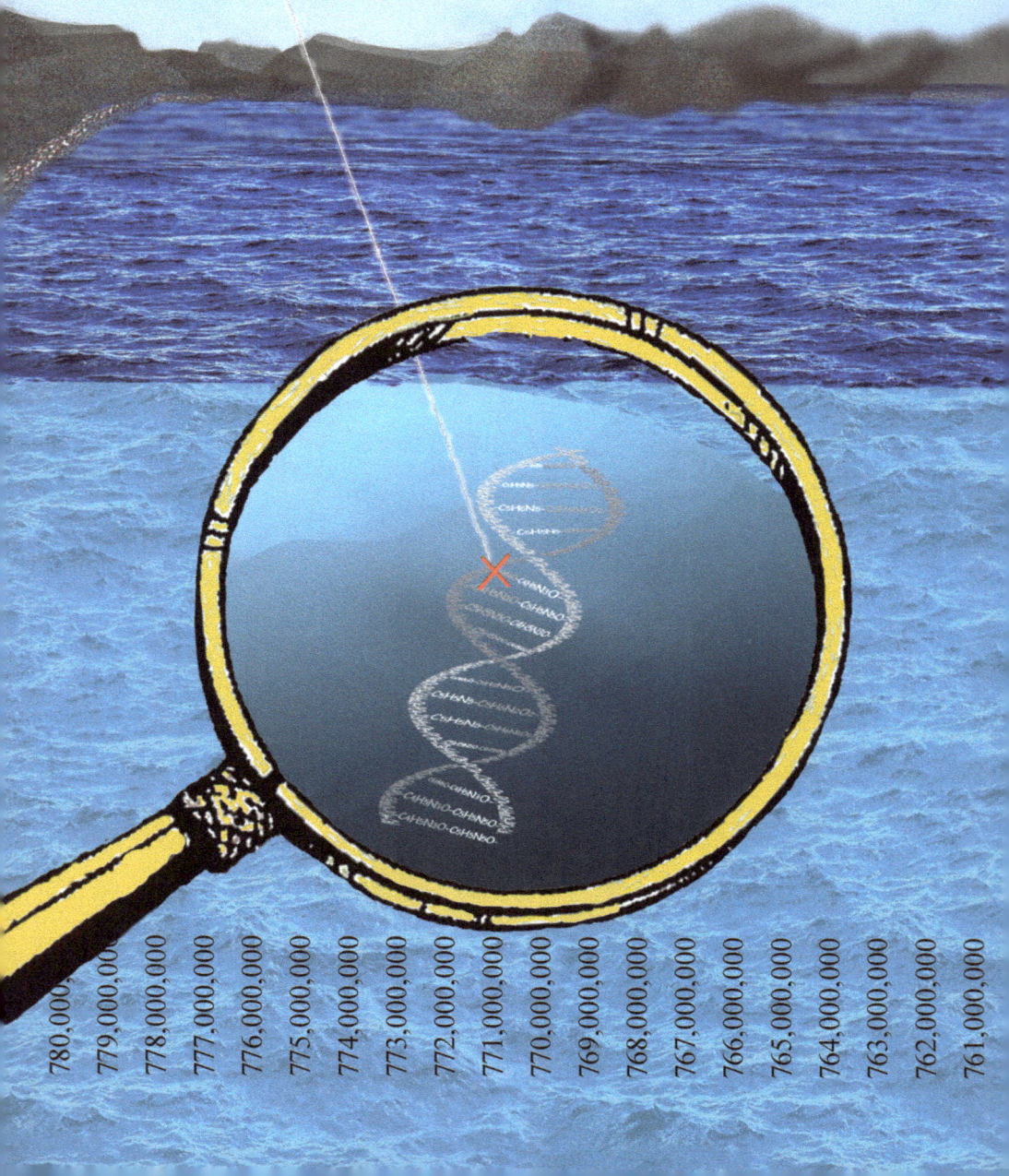

9. FORMATION OF THE OZONE LAYER

Now, UV radiation from the Sun
has long impeded life's full flourishing,
but thanks to some two billion steady years
of photosynthesis, the time now comes
to benefit from yet another change:

O₂

760,000,000
759,000,000
758,000,000
757,000,000
756,000,000
755,000,000
754,000,000
753,000,000
752,000,000
751,000,000
750,000,000
749,000,000
748,000,000
747,000,000
746,000,000
745,000,000
744,000,000
743,000,000
742,000,000
741,000,000

O_2
O_2
O_2
O_2
O_2

O_2
O_2
O_2

O_3
O_3 O_3
O_3 O_3 O_3
O_3
O_3
O_3

$O_2 \rightarrow$

O
O

$\rightarrow O_3$

O_2

O_2
O_2
O_2
O_2
O_2

O_2
O_2
O_2
O_2
O_2

O_2
O_2
O_2
O_2

740,000,000
739,000,000
738,000,000
737,000,000
736,000,000
735,000,000
734,000,000
733,000,000
732,000,000
731,000,000
730,000,000
729,000,000
728,000,000
727,000,000
726,000,000
725,000,000
724,000,000
723,000,000
722,000,000
721,000,000

High in the atmosphere, when split apart
by ultraviolet rays, some oxygen
is turned into an ozone shield, which now
protects the workings of our DNA,
and free to replicate without the curse
of UV radiation, we're now free
to explore newfound opportunities.

The ozone layer begins forming about 700 mya

720,000,000 719,000,000 718,000,000 717,000,000 716,000,000 715,000,000 714,000,000 713,000,000 712,000,000 711,000,000 710,000,000 709,000,000 708,000,000 707,000,000 706,000,000 705,000,000 704,000,000 703,000,000 702,000,000 701,000,000

10. EXPLOSION

And oh, what opportunities there are!
Protected from those harmful UV rays,
we're capable of doing anything!
We colonize the land with chlorophyll
and paint the landscape green with verdant life.

O₂ O₂ O₂ O₂

700,000,000 699,000,000 698,000,000 697,000,000 696,000,000 695,000,000 694,000,000 693,000,000 692,000,000 691,000,000 690,000,000 689,000,000 688,000,000 687,000,000 686,000,000 685,000,000 684,000,000 683,000,000 682,000,000 681,000,000

We turn into fierce protozoans, beasts
that hunt for food as predators on prey,
consuming other cells to feed upon
their stores of precious, built-up energy.

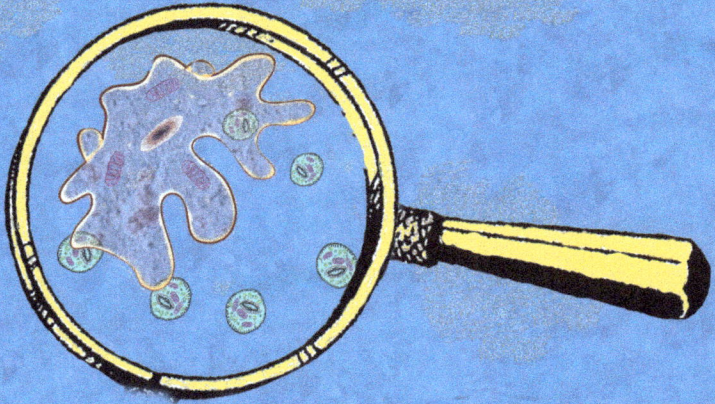

680,000,000
679,000,000
678,000,000
677,000,000
676,000,000
675,000,000
674,000,000
673,000,000
672,000,000
671,000,000
670,000,000
669,000,000
668,000,000
667,000,000
666,000,000
665,000,000
664,000,000
663,000,000
662,000,000
661,000,000

As oxygen replaces CO_2
we who obtain our fuel from it prosper,
and as we thrive, bacteria consume
our carcasses, turning decay into
nutritious fields of soil, and viruses
find more host cells in which to replicate.
The Earth and life evolve as if one thing.

O_2 O_2 O_2 O_2 O_2 O_2 O_2 O_2 O_2 O_2 O_2 O_2 O_2 O_2 O_2 O_2 O_2 O_2 O_2 O_2 O_2 O_2 O_2 O_2 O_2 O_2 O_2 O_2 O_2 O_2 O_2 O_2

660,000,000 659,000,000 658,000,000 657,000,000 656,000,000 655,000,000 654,000,000 653,000,000 652,000,000 651,000,000 650,000,000 649,000,000 648,000,000 647,000,000 646,000,000 645,000,000 644,000,000 643,000,000 642,000,000 641,000,000

Now lower atmospheric CO_2
leaves less to capture sunlight's precious warmth,
and temperatures on Earth begin to fall
until the whole world freezes over.

640,000,000
639,000,000
638,000,000
637,000,000
636,000,000
635,000,000
634,000,000
633,000,000
632,000,000
631,000,000
630,000,000
629,000,000
628,000,000
627,000,000
626,000,000
625,000,000
624,000,000
623,000,000
622,000,000
621,000,000

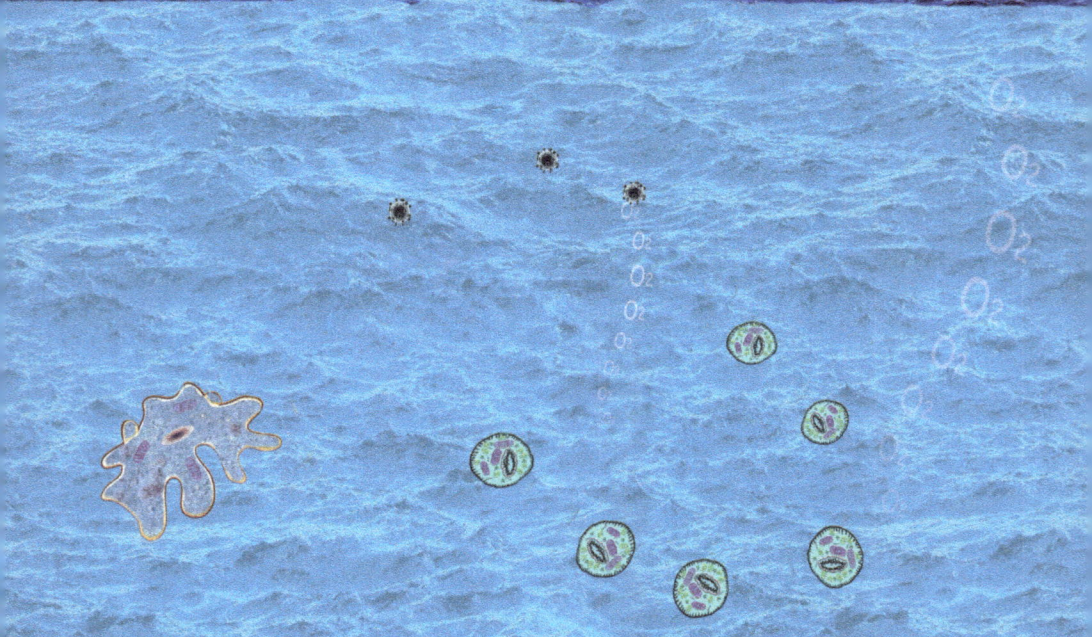

620,000,000
619,000,000
618,000,000
617,000,000
616,000,000
615,000,000
614,000,000
613,000,000
612,000,000
611,000,000
610,000,000
609,000,000
608,000,000
607,000,000
606,000,000
605,000,000
604,000,000
603,000,000
602,000,000
601,000,000

O_2
O_2
O_2
O_2

Still,
the viruses, bacteria, and most
eukaryotes do fine, because the time
we have to acclimate to colder days
is fifty million years or so.

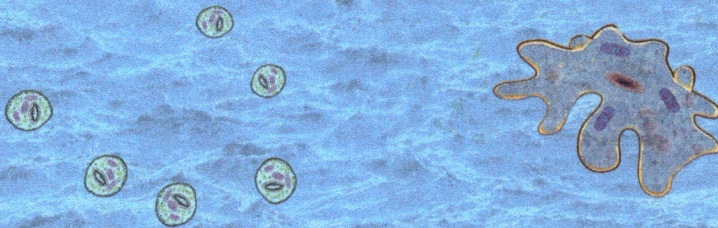

600,000,000
599,000,000
598,000,000
597,000,000
596,000,000
595,000,000
594,000,000
593,000,000
592,000,000
591,000,000
590,000,000
589,000,000
588,000,000
587,000,000
586,000,000
585,000,000
584,000,000
583,000,000
582,000,000
581,000,000

And when
activity inside Earth's core creates
a surge in volcanic eruptions all
around the globe, releasing CO_2
and warming up the atmosphere again,
the story is the same: all life adapts
once more, reverting to our warmer ways,
for simple things have all the time we need
to make new genes when faced with climate change.

580,000,000
579,000,000
578,000,000
577,000,000
576,000,000
575,000,000
574,000,000
573,000,000
572,000,000
571,000,000
570,000,000
569,000,000
568,000,000
567,000,000
566,000,000
56
563,000,000
562,000,000
561,000,000

CO_2
CO_2
CO_2
CO_2

But multi-celled eukaryotes are not
such simple things. As each new type of cell
develops its peculiar specialty,
we grow into all sorts of complex forms,
like jellyfish with sprawling tentacles
and bilateria with heads and tails.

560,000,000
559,000,000
558,000,000
557,000,000
556,000,000
555,000,000
554,000,000
553,000,000
552,000,000
551,000,000
550,000,000
549,000,000
548,000,000
547,000,000
546,000,000
545,000,000
544,000,000
543,000,000
542,000,000
541,000,000

We grow into stiff-legged arthropods
with paired appendages, segmented worms
with mouths in front and pairs of eyes in back,
and creatures housed in shells, and trilobites
with triple lobes, and vertebrates with spines,
and barnacles, and sponges, and much more.

The "Cambrian Explosion" of complex forms of life

540,000,000
539,000,000
538,000,000
537,000,000
536,000,000
535,000,000
534,000,000
533,000,000
532,000,000
531,000,000
530,000,000
529,000,000
528,000,000
527,000,000
526,000,000
525,000,000
524,000,000
523,000,000
522,000,000
521,000,000

And such complexity comes with a price.
With such variety of parts, each part
dependent on the others to survive,
our DNA must keep it all just right —
genetic change becomes most cumbersome.
So when another round of volcanoes
sends CO_2 into the sky again,
and days get warm and seas acidify,
great species that have lived ten million years
cannot adapt quite fast enough — and die.

End-Botomian Extinction Event 517 mya: 40% of marine genera become extinct

Dresbachian Extinction Event 502 mya
40% of marine genera become extinct

520,000,000
519,000,000
518,000,000
517,000,000
516,000,000
515,000,000
514,000,000
513,000,000
512,000,000
511,000,000
510,000,000
509,000,000
508,000,000
507,000,000
506,000,000
505,000,000
504,000,000
503,000,000
502,000,000
501,000,000

For several million years, green algae thrive,
adjusting to what's new, but yet again
volcanoes spew out CO_2; the air
gets full of greenhouse gas; the ocean warms
as oxygen declines. The simplest things
adapt, but complex forms need too much time.
The trilobites, the sharp-toothed conodonts,
brachiopods with hinged bi-valves like clams,
are so adapted to things as they are,
that when change comes, whole orders of them die.

Cambrian-Ordovician Extinction Event
485 mya, wipes out many species of
brachiopods, conodonts and trilobites.

500,000,000
499,000,000
498,000,000
497,000,000
496,000,000
495,000,000
494,000,000
493,000,000
492,000,000
491,000,000
490,000,000
489,000,000
488,000,000
487,000,000
486,000,000
485,000,000
484,000,000
483,000,000
482,000,000
481,000,000

Some forms of life recover. Simpler things
fill niches now vacated by those gone,
and though it takes us many, many years,
we now see starfish on the ocean floor,
vast coral beds arise, and sure enough,
we colonize the land again, this time
with mosses, fungi, and green liverworts.

481,000,000
479,000,000
478,000,000
477,000,000
476,000,000
475,000,000
474,000,000
473,000,000
472,000,000
471,000,000
470,000,000
469,000,000
468,000,000
467,000,000
466,000,000
465,000,000
464,000,000
463,000,000
462,000,000
461,000,000

Now, some say gamma rays from outer space destroy our treasured ozone shield, and some say unchecked photosynthesis has cleared the air of too much CO_2 — what's clear is that a drop in greenhouse gases now sends Earth into a freeze, and eighty-five percent of ocean life does not survive.

End-Ordovician Extinction Event 440 myc wipes out many brachiopods, conodonts echinoderms, graptolites and trilobites.

→

460,000,000 459,000,000 458,000,000 457,000,000 456,000,000 455,000,000 454,000,000 453,000,000 452,000,000 451,000,000 450,000,000 449,000,000 448,000,000 447,000,000 446,000,000 445,000,000 444,000,000 443,000,000 442,000,000 441,000,000

A hundred thousand years go by, and then
a million more. Those who withstood the freeze
now fill the niches left by those who died.
This time, the mollusks thrive, and bony fish
with shark-like jaws, and on dry land, new plants
with special tissues to transport their food.

Whole families of jawless fish decline
as those with jaws advance. And fish with fins
come to the fore, some made with fleshy lobes
and some with radiating bones — and some
so hard and tough that in some shallow pools
they push against the sea floor to propel
their bodies on.

420,000,000
419,000,000
418,000,000
417,000,000
416,000,000
415,000,000
414,000,000
413,000,000
412,000,000
411,000,000
410,000,000
409,000,000
408,000,000
407,000,000
406,000,000
405,000,000
404,000,000
403,000,000
402,000,000
401,000,000

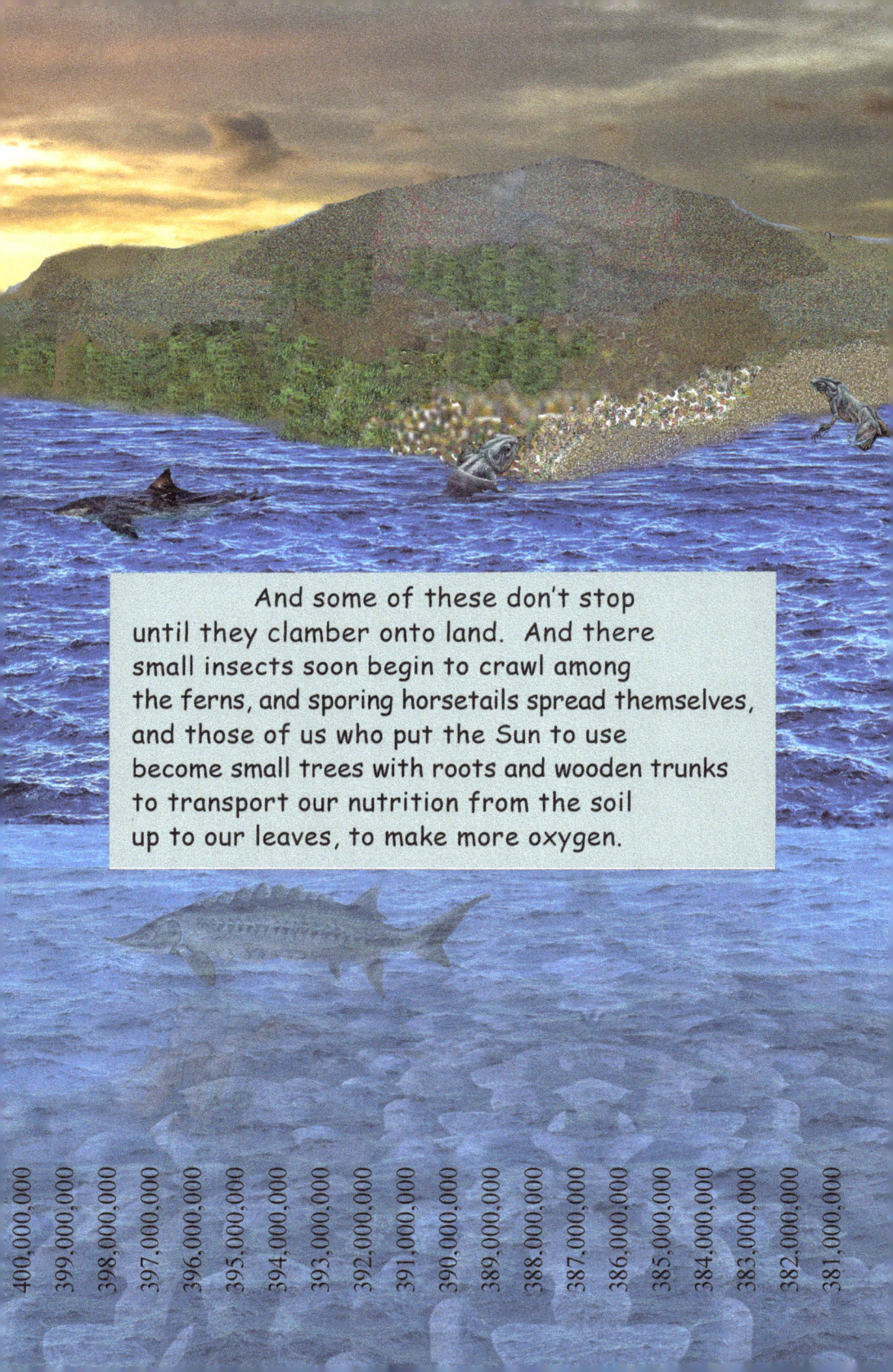

And some of these don't stop
until they clamber onto land. And there
small insects soon begin to crawl among
the ferns, and sporing horsetails spread themselves,
and those of us who put the Sun to use
become small trees with roots and wooden trunks
to transport our nutrition from the soil
up to our leaves, to make more oxygen.

400,000,000 399,000,000 398,000,000 397,000,000 396,000,000 395,000,000 394,000,000 393,000,000 392,000,000 391,000,000 390,000,000 389,000,000 388,000,000 387,000,000 386,000,000 385,000,000 384,000,000 383,000,000 382,000,000 381,000,000

Some fern-like trees with fronds grow eighty feet and others produce seeds, as arthropods walk nimbly on dry land and animals with backbones multiply. Yet changes in the ocean's temperature and decreases in oxygen continue to mean death for complex, multi-celled eukaryotes.

Late Devonian Extinctions, 375-360 mya, wipe out the jawless fishes, the elpistostegelians and most trilobites.

380,000,000 379,000,000 378,000,000 377,000,000 376,000,000 375,000,000 374,000,000 373,000,000 372,000,000 371,000,000 370,000,000 369,000,000 368,000,000 367,000,000 366,000,000 365,000,000 364,000,000 363,000,000 362,000,000 361,000,000

Now, where successful forms of life decline,
the void they leave is filled by other things.
so mollusks like the ammonites survive,
and tetrapods move back and forth between
the water and dry land with gill-like lungs.
Soon, early reptiles form, and insects learn
to fly (like dragonflies with two foot wings).

The continents now join as Pangaea, whose mountain peaks and deep ravines collect and carry rain down to the sea. The air grows hot and humid as the trees create a vast expanse of living greenery.

340,000,000
339,000,000
338,000,000
337,000,000
336,000,000
335,000,000
334,000,000
333,000,000
332,000,000
331,000,000
330,000,000
329,000,000
328,000,000
327,000,000
326,000,000
325,000,000
324,000,000
323,000,000
322,000,000
321,000,000

For more or less a hundred million years, amphibians have their heyday in the swamps and fertile pools beneath the canopy. But gradually, the climate cools again, evolving from the hot and humid world to one that's dry. And with the loss of rain comes tragedy: the rainforests collapse.

Carboniferous Rainforest Collapse 305 mya, wipes out many species. Amphibians lose their dominance.

320.000.000
319.000.000
318.000.000
317.000.000
316.000.000
315.000.000
314.000.000
313.000.000
312.000.000
311.000.000
310.000.000
309.000.000
308.000.000
307.000.000
306.000.000
305.000.000
304.000.000
303.000.000
302.000.000
301.000.000

With increased CO_2 and oxygen from forests now reduced, the land grows hot and dry; amphibians, who lay their eggs in water, can't compete with animals who carry young in amniotic sacs — the synapsids and saurapsids who come to rule dry land in this, the Permian age.

300,000,000 299,000,000 298,000,000 297,000,000 296,000,000 295,000,000 294,000,000 293,000,000 292,000,000 291,000,000 290,000,000 289,000,000 288,000,000 287,000,000 286,000,000 285,000,000 284,000,000 283,000,000 282,000,000 281,000,000

Bacteria and viruses survive.
And in the tens of million years that lack
of water is a threat to life on land,
the arid climate proves a welcome boon
to insects with their tiny shells, and those
who travel fast and far, from water holes
to feeding grounds, by learning how to fly.

280,000,000 279,000,000 278,000,000 277,000,000 276,000,000 275,000,000 274,000,000 273,000,000 272,000,000 271,000,000 270,000,000 269,000,000 268,000,000 267,000,000 266,000,000 265,000,000 264,000,000 263,000,000 262,000,000 261,000,000

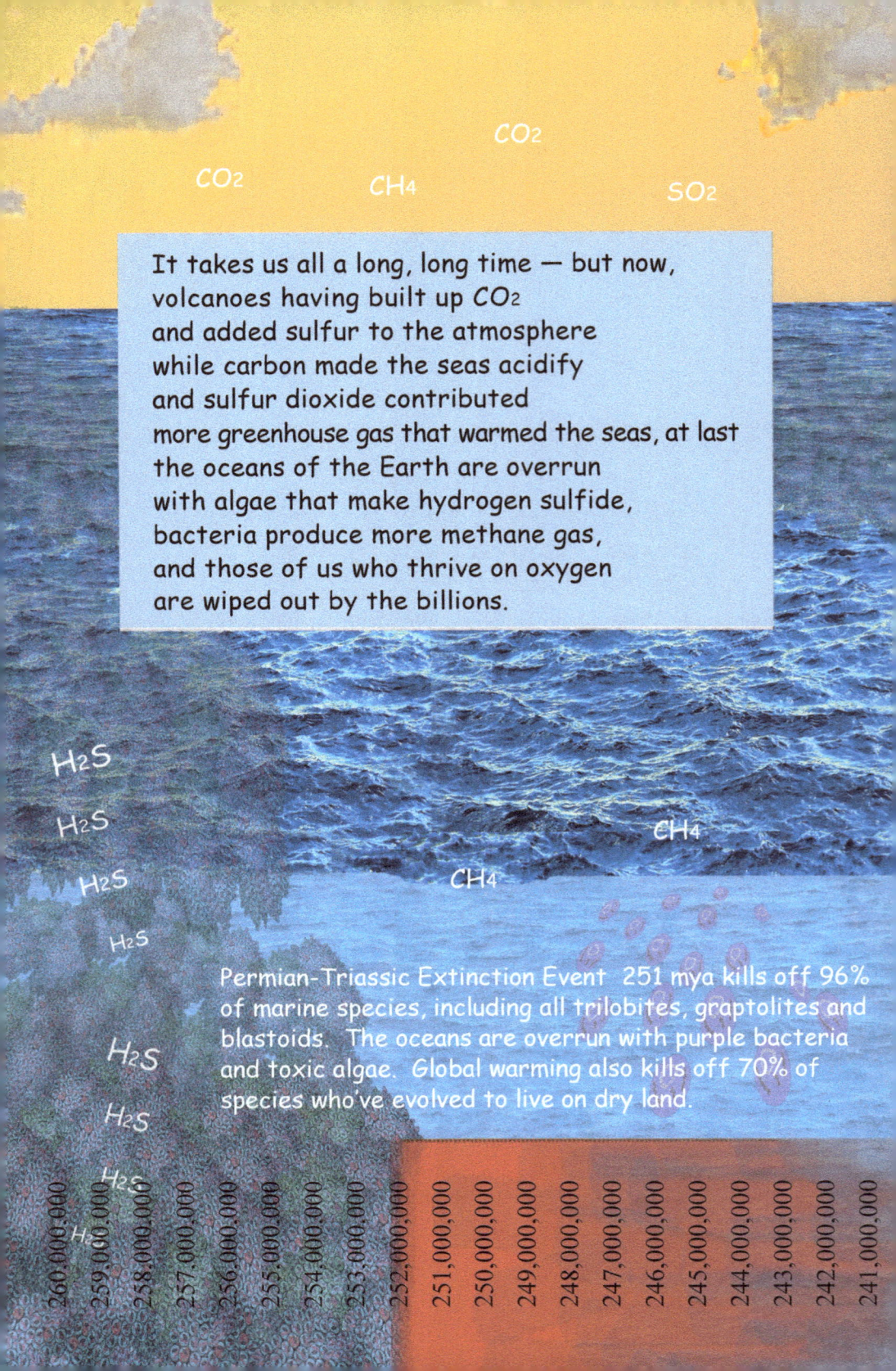

CO_2

CO_2 CH_4 SO_2

It takes us all a long, long time — but now,
volcanoes having built up CO_2
and added sulfur to the atmosphere
while carbon made the seas acidify
and sulfur dioxide contributed
more greenhouse gas that warmed the seas, at last
the oceans of the Earth are overrun
with algae that make hydrogen sulfide,
bacteria produce more methane gas,
and those of us who thrive on oxygen
are wiped out by the billions.

H_2S

H_2S

H_2S CH_4

H_2S CH_4

Permian-Triassic Extinction Event 251 mya kills off 96%
of marine species, including all trilobites, graptolites and
blastoids. The oceans are overrun with purple bacteria
H_2S and toxic algae. Global warming also kills off 70% of
species who've evolved to live on dry land.

H_2S

H_2S

H_2

260,000,000 | 259,000,000 | 258,000,000 | 257,000,000 | 256,000,000 | 255,000,000 | 254,000,000 | 253,000,000 | 252,000,000 | 251,000,000 | 250,000,000 | 249,000,000 | 248,000,000 | 247,000,000 | 246,000,000 | 245,000,000 | 244,000,000 | 243,000,000 | 242,000,000 | 241,000,000

Still, despite
the devastating blow to some, it is
their loss that clears the stage for others who
adapt to Earth's new climate. Seas are filled
with urchins, slimy snails and scuttling crabs,
icthyosaurs and early crocodiles.
The land gives birth to conifers with cones
and beasts called therapsids, whose limbs aren't splayed
out to their sides, but set beneath their trunks.

240,000,000 239,000,000 238,000,000 237,000,000 236,000,000 235,000,000 234,000,000 233,000,000 232,000,000 231,000,000 230,000,000 229,000,000 228,000,000 227,000,000 226,000,000 225,000,000 224,000,000 223,000,000 222,000,000 1,000,000

Among them, armor-plated aetosaurs
arise and walk the Earth, while crocodiles
now gather nests in which to lay their eggs.
But miles away, eruptions once again
release large quantities of CO_2
that warm the air, acidify the sea,
and lead to mass extinctions everywhere.

Triassic-Jurassic Extinction Event culminating
201 mya sees the demise of 76% of all species
on land and sea, including disappearance of all
the conodonts and aetosaurs and 60% of all
pollen-bearing plants.

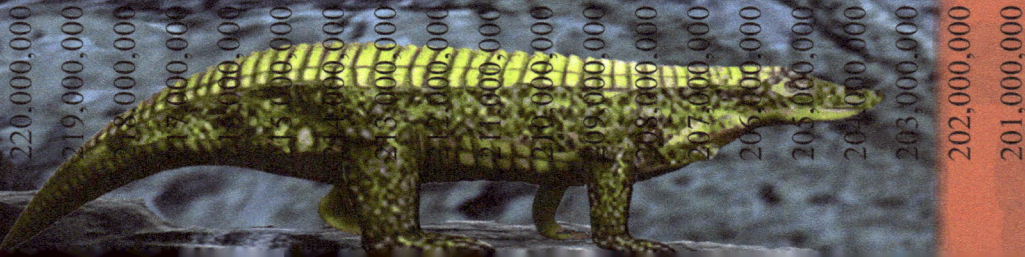

220,000,000 219,000,000 218,000,000 217,000,000 216,000,000 215,000,000 214,000,000 213,000,000 212,000,000 211,000,000 210,000,000 209,000,000 208,000,000 207,000,000 206,000,000 205,000,000 204,000,000 203,000,000 202,000,000 201,000,000

So now the dinosaurs take center stage.
At thirty tons and thirty meters long,
huge sauropods now dominate the Earth,
while little lizards, salamanders, frogs
and sea urchins become more plentiful.

200,000,000
199,000,000
198,000,000
197,000,000
196,000,000
195,000,000
194,000,000
193,000,000
192,000,000
191,000,000
190,000,000
189,000,000
188,000,000
187,000,000
186,000,000
185,000,000
184,000,000
183,000,000
182,000,000
181,000,000

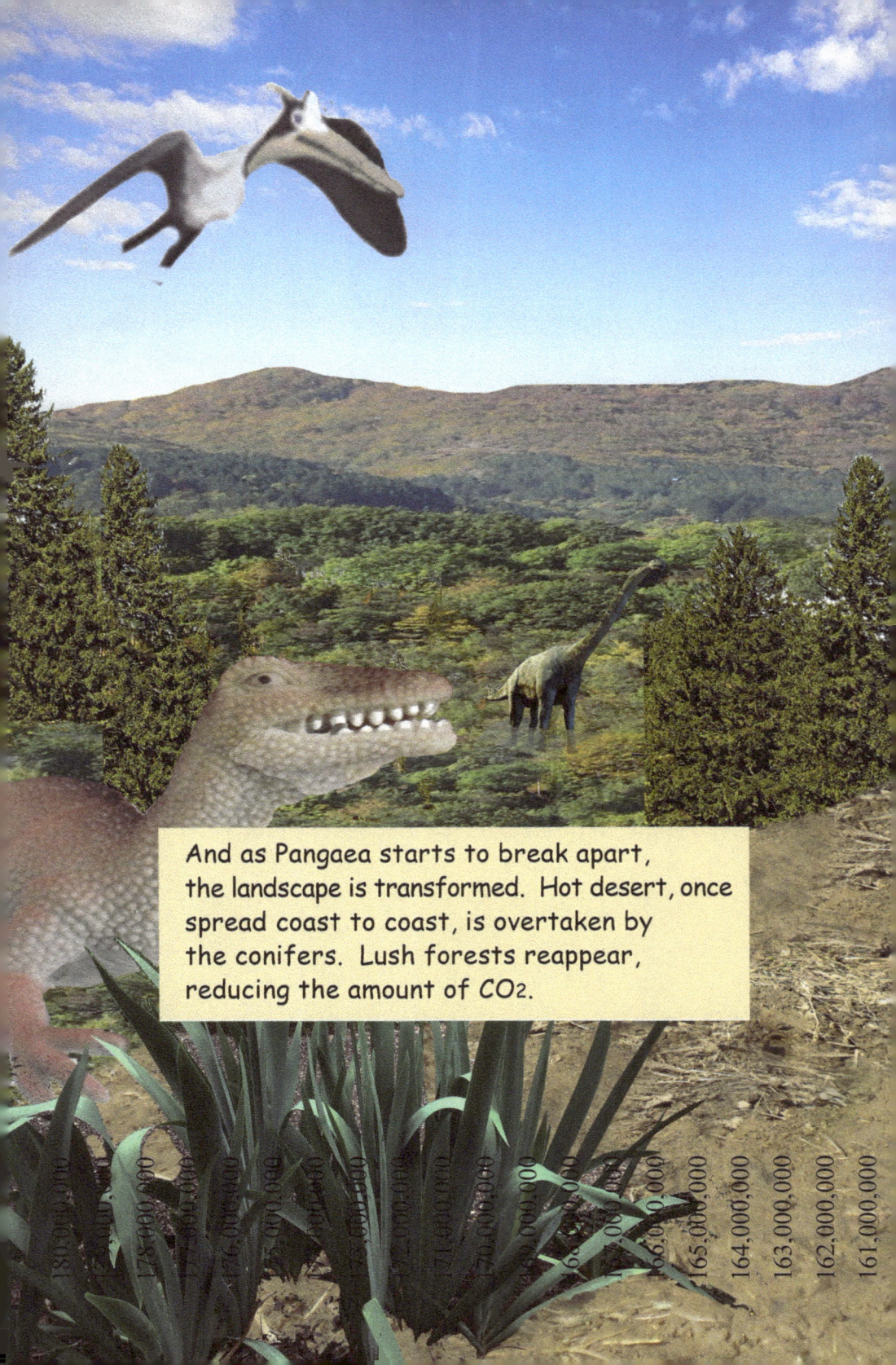

And as Pangaea starts to break apart, the landscape is transformed. Hot desert, once spread coast to coast, is overtaken by the conifers. Lush forests reappear, reducing the amount of CO_2.

180,000,000 178,000,000 176,000,000 175,000,000 173,000,000 172,000,000 171,000,000 169,000,000 167,000,000 166,000,000 165,000,000 164,000,000 163,000,000 162,000,000 161,000,000

Among the modest creatures, not so tall
as sauropods, are little animals
with bigger brains and better sense of smell
that burrow in the earth during the day,
emerging after dark (when carnivores
have gone to sleep) to hunt for bugs to eat.

160,000,000 159,000,000 158,000,000 157,000,000 156,000,000 155,000,000 154,000,000 153,000,000 152,000,000 151,000,000 150,000,000 149,000,000 148,000,000 147,000,000 146,000,000 145,000,000 144,000,000 143,000,000 142,000,000 141,000,000

Eukaryotic life takes every size
and shape. At sea, the hard-shelled ammonites
join fish with gills and scales while, inland, plants
begin to flower, insects pollinate.

140,000,000   139,000,000   138,000,000   137,000,000   136,000,000   135,000,000   134,000,000   133,000,000   132,000,000   131,000,000   130,000,000   129,000,000   128,000,000   127,000,000   126,000,000   125,000,000   124,000,000   123,000,000   122,000,000   121,000,000

And while the reptiles hatch their young from eggs,
the furry little animals who sneak
out every night, evading predators,
develop a placenta that can feed
an embryonic life, and pelvic bones
that separate when birth is close at hand.

120,000,000
119,000,000
118,000,000
117,000,000
116,000,000
115,000,000
114,000,000
113,000,000
112,000,000
111,000,000
110,000,000
109,000,000
108,000,000
107,000,000
106,000,000
105,000,000
104,000,000
103,000,000
102,000,000
101,000,000

The dinosaurs — those apex predators exuding confidence with every breath — have reigned a hundred fifty million years above all other things, so might have cause to think themselves invincible to dust and tiny molecules like CO_2.

100,000,000 99,000,000 98,000,000 97,000,000 96,000,000 95,000,000 94,000,000 93,000,000 92,000,000 91,000,000 90,000,000 89,000,000 88,000,000 87,000,000 86,000,000 85,000,000 84,000,000 83,000,000 82,000,000 81,000,000

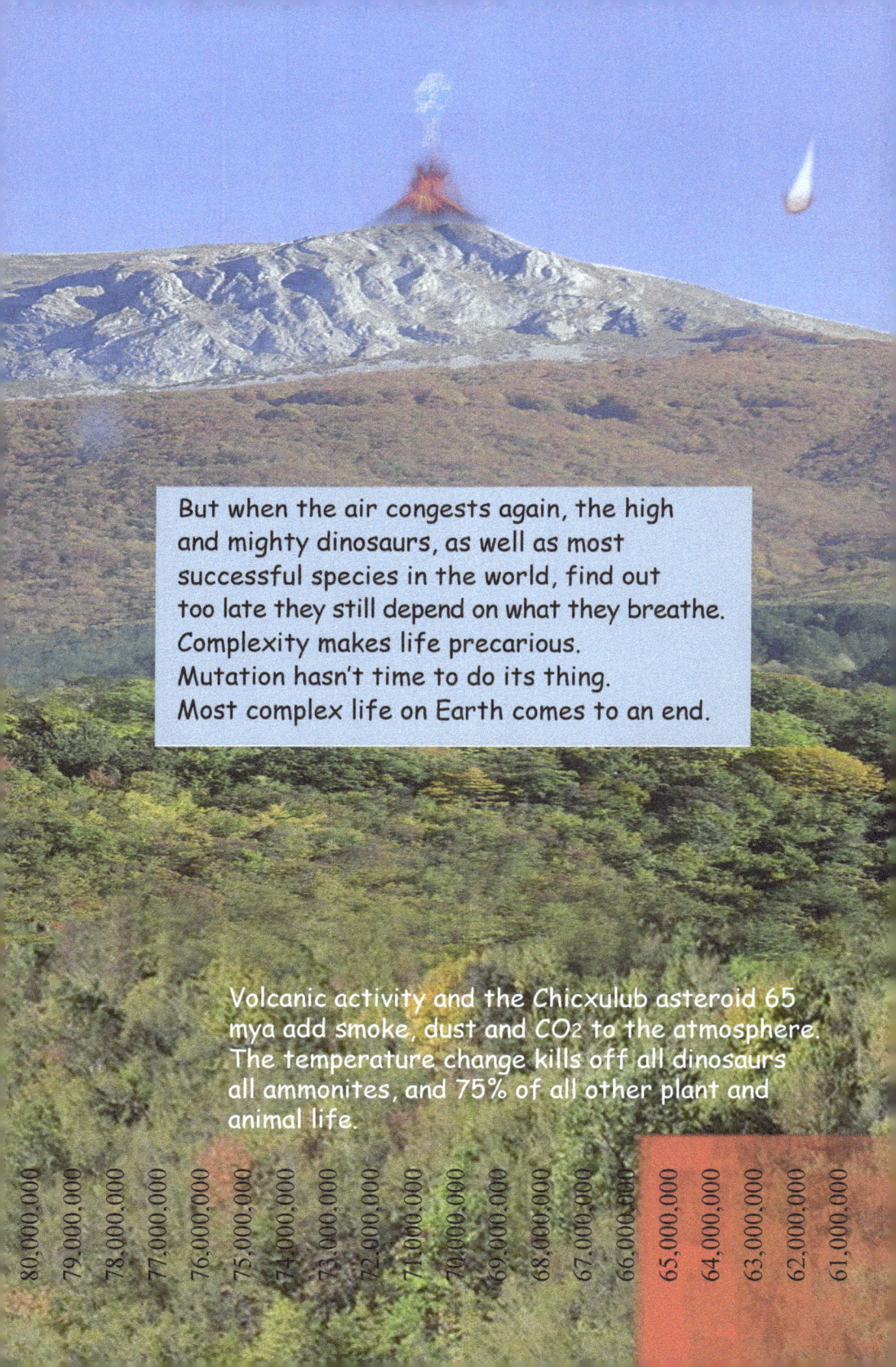

But when the air congests again, the high
and mighty dinosaurs, as well as most
successful species in the world, find out
too late they still depend on what they breathe.
Complexity makes life precarious.
Mutation hasn't time to do its thing.
Most complex life on Earth comes to an end.

Volcanic activity and the Chicxulub asteroid 65
mya add smoke, dust and CO_2 to the atmosphere.
The temperature change kills off all dinosaurs
all ammonites, and 75% of all other plant and
animal life.

80,000,000 79,000,000 78,000,000 77,000,000 76,000,000 75,000,000 74,000,000 73,000,000 72,000,000 71,000,000 70,000,000 69,000,000 68,000,000 67,000,000 66,000,000 65,000,000 64,000,000 63,000,000 62,000,000 61,000,000

But death to mighty things has benefits.
In time, the trees restore the oxygen.
The clearing of the atmosphere creates
a world in which the feathered birds can fly,
in which large mammals can produce their young
and nurse them by the light of day, and some
can even make safe homes up in the trees.

60,000,000 59,000,000 58,000,000 57,000,000 56,000,000 55,000,000 54,000,000 53,000,000 52,000,000 51,000,000 50,000,000 49,000,000 48,000,000 47,000,000 46,000,000 45,000,000 44,000,000 43,000,000 42,000,000 41,000,000

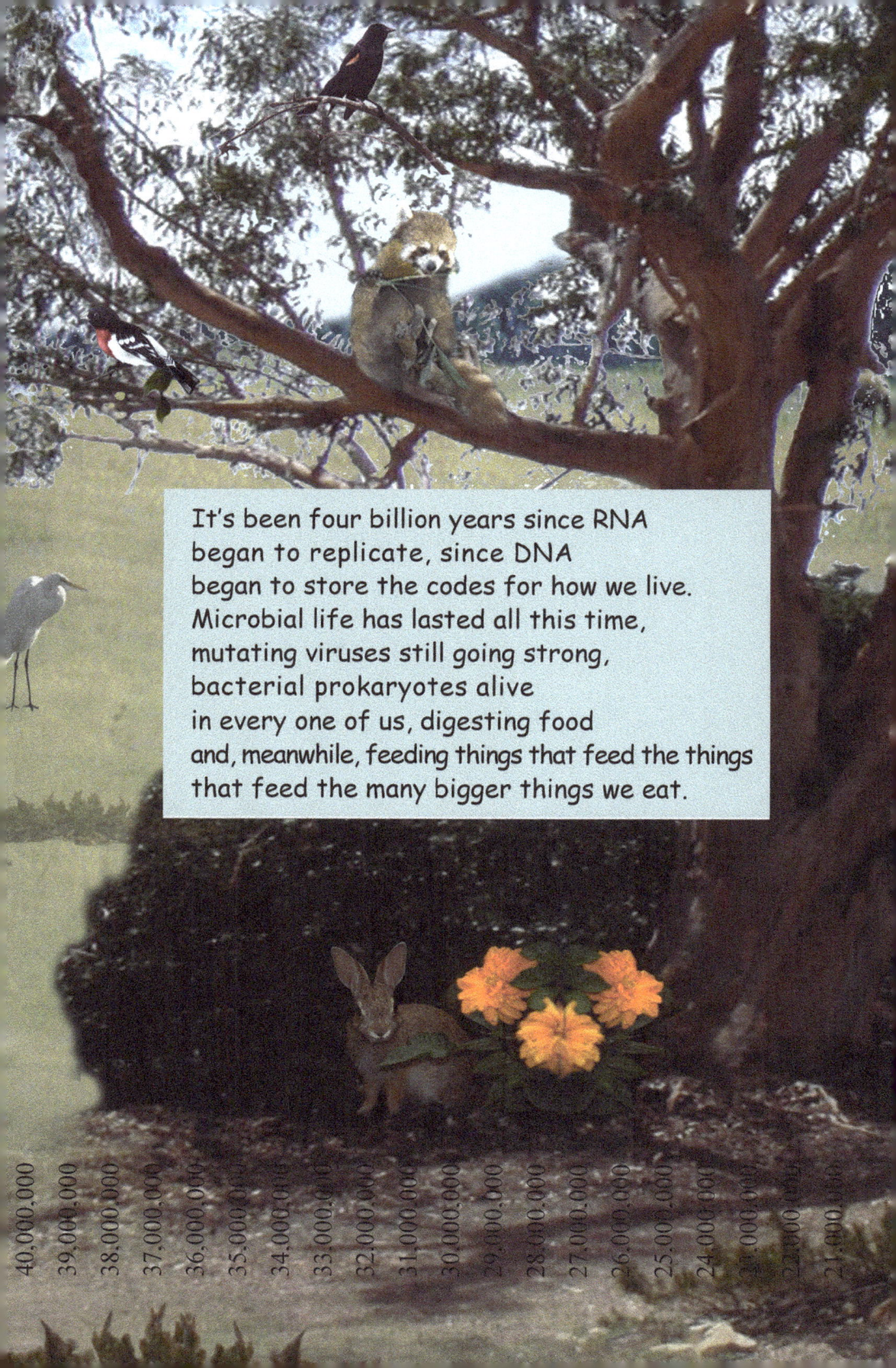

It's been four billion years since RNA
began to replicate, since DNA
began to store the codes for how we live.
Microbial life has lasted all this time,
mutating viruses still going strong,
bacterial prokaryotes alive
in every one of us, digesting food
and, meanwhile, feeding things that feed the things
that feed the many bigger things we eat.

40,000,000
39,000,000
38,000,000
37,000,000
36,000,000
35,000,000
34,000,000
33,000,000
32,000,000
31,000,000
30,000,000
29,000,000
28,000,000
27,000,000
26,000,000
25,000,000
24,000,000
23,000,000
22,000,000
21,000,000

After four billion years of life on Earth,
the age has come of whales and mastodons,
of polar bears, great apes and dodo birds:
all manner of complex eukaryotes,
dependent on the world that we all share.

      *   *   *

Befuddled by small things we cannot see,
some of us gaze upon the vast blue sky
believing that the sun, the moon, and all
the creatures of the Earth were made for us

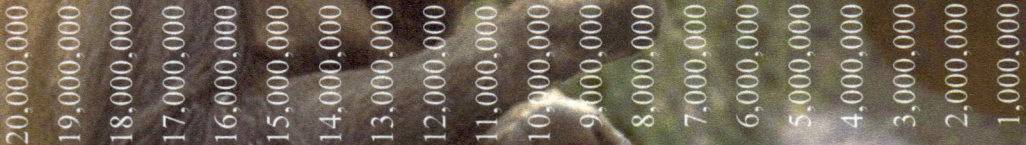

Apes begin to walk upright
North and South
America connect

20,000,000   19,000,000   18,000,000   17,000,000   16,000,000   15,000,000   14,000,000   13,000,000   12,000,000   11,000,000   10,000,000   9,000,000   8,000,000   7,000,000   6,000,000   5,000,000   4,000,000   3,000,000   2,000,000   1,000,000

THE LAST MILLION YEARS

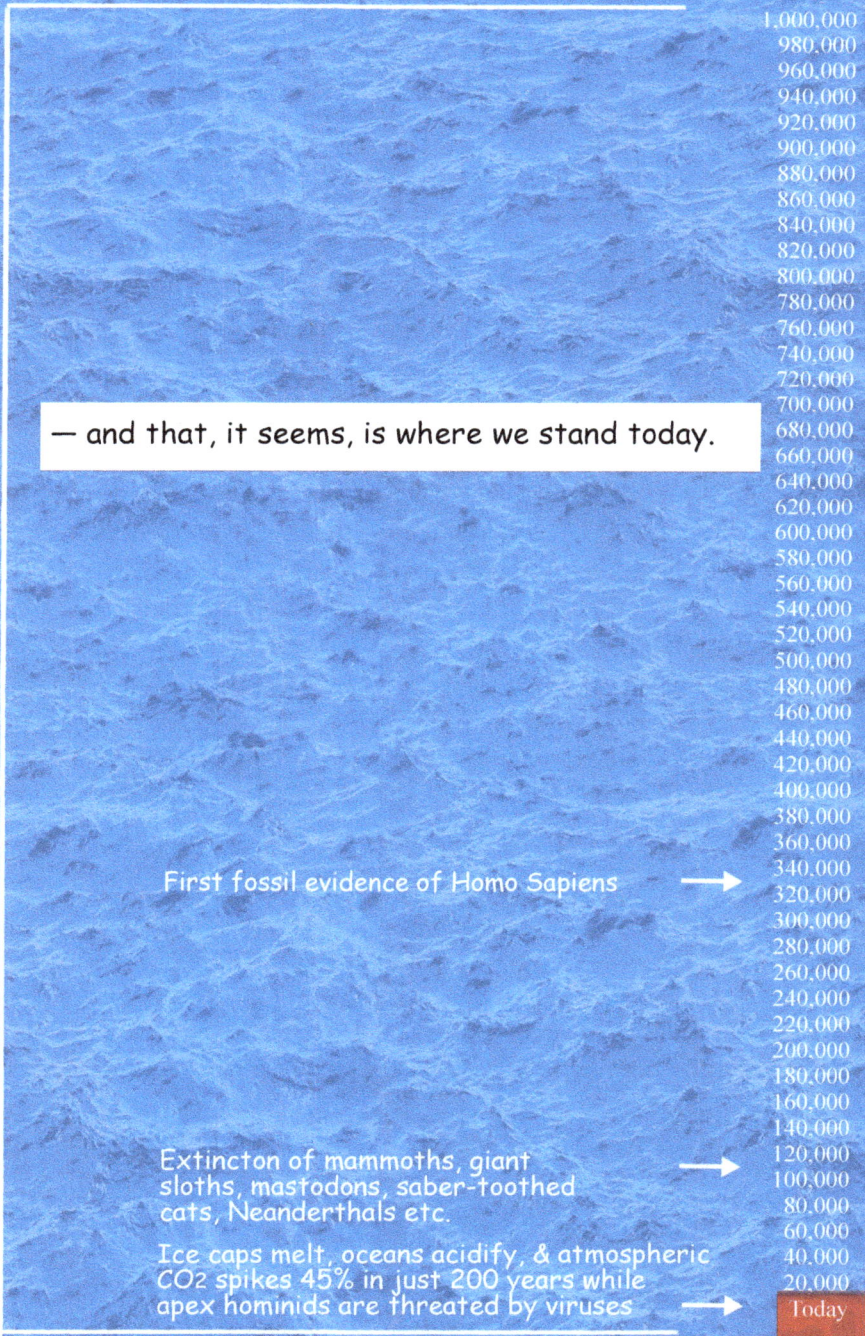

1,000,000
980,000
960,000
940,000
920,000
900,000
880,000
860,000
840,000
820,000
800,000
780,000
760,000
740,000
720,000
700,000
680,000
660,000
640,000
620,000
600,000
580,000
560,000
540,000
520,000
500,000
480,000
460,000
440,000
420,000
400,000
380,000
360,000
340,000
320,000
300,000
280,000
260,000
240,000
220,000
200,000
180,000
160,000
140,000
120,000
100,000
80,000
60,000
40,000
20,000
Today

— and that, it seems, is where we stand today.

First fossil evidence of Homo Sapiens →

Extincton of mammoths, giant sloths, mastodons, saber-toothed cats, Neanderthals etc. →

Ice caps melt, oceans acidify, & atmospheric CO_2 spikes 45% in just 200 years while apex hominids are threated by viruses →

Suggested Reading

The history of life
I've tried to express in images and
blank verse represents what I've gleaned
from the works of learned scholars and
authors, to whom I am most grateful.
Those I've found most fascinating
include

*Life on a Young Planet: The First Three Billion Years
of Evolution on Earth*
by Andrew Knoll (Princeton Univ. Press, 2003)

Maps of Time: An Introduction to Big History
by David Christian (Univ. of California, 2004)

The Sixth Extinction: An Unnatural History
by Elizabeth Kolbert (Henry Holt & Co., 2014)

A New History of Life
by Peter Ward and Joe Kirschvink (Bloomsbury, 2015)

The Vital Question: Why Is Life the Way It Is?
by Nick Lane (Profile Books, 2015)

The Tangled Tree: A Radical New History of Life
by David Quammen (Simon & Schuster, 2018)

*Nothing
in
Common*

About the Author

Joseph W. Carvin writes *WeMayBeWrong*, a blog devoted to the connections between fallibility, humility, and civility. He currently lives in Virginia with his wife Karen, two of their four children, five of their six grandchildren, a cat, two birds, two lizards, two hamsters, four dogs, eleven chickens, countless bees and butterflies, and many millions of prokaryotes — all of whom help make his life a great joy.

Previous books:

Cage Stories: Memories of Fatherhood and Creation,
ISBN 978-0-9768183-1-1

A Piece of the Pie: The Story of Customer Service at Publix,
ISBN 978-0-9768183-7-3
ISBN 978-1-7332515-0-1

Oh, Mother, That Man's Here Again!!
The Christmas Cards of Charles W. Carvin,
ISBN 978-0-9768183-5-9

Alemeth
ISBN 978-0-9768183-8-0

Praise for Alemeth:
Carvin masterfully brings to life a South in dramatic transition, and he avoids the binary categories of pro and con that often typify the genre... [A] thoughtful, sensitive rendering of a complex period in American history. A philosophically challenging look at the inner turmoil of the American South in the 19th century.
— *Kirkus Reviews*

The author can be contacted on the web at jwcarvin.com or *wemaybewrong.org*, or by e-mail at *jwcarvin@gmail.com*.

www.ingramcontent.com/pod-product-compliance
Lightning Source LLC
Chambersburg PA
CBHW040921210326
41597CB00030B/5146